DEAD MATTER

A RINEHART SUSPENSE NOVEL

A RINEHART SUSPENSE NOVEL

DEAD MATTER

Steven Frimmer

HOLT, RINEHART AND WINSTON
New York

Copyright © 1982 by Steven Frimmer
All rights reserved, including the right to reproduce this
book or portions thereof in any form.
Published by Holt, Rinehart and Winston,
383 Madison Avenue, New York, New York 10017.
Published simultaneously in Canada by Holt, Rinehart
and Winston of Canada, Limited.

Library of Congress Cataloging in Publication Data

Frimmer, Steven.
Dead matter.
(A Rinehart suspense novel)
I. Title.
PS3556.R5687D4   813'.54   81-6535
AACR2
ISBN: 0-03-059114-7

First Edition

DESIGNER: Lucy Castelluccio
Printed in the United States of America
1 3 5 7 9 10 8 6 4 2

While this is a work of fiction, I have drawn upon my recollections of actual incidents and situations that occurred in and out of publishing. I have also combined attributes of real individuals into the composites embodied by the various characters in this book. Therefore, if my friends in publishing think they see themselves or others known to them in some of these characters, they are, at best, only partly correct.

ISBN 0-03-059114-7

This book is for my wife, Barbara, who has often asked what it is an editor does all day.

DEAD MATTER

A RINEHART SUSPENSE NOVEL

Most ardent New Yorkers come from someplace else. The born-and-bred New Yorkers have largely moved to the suburbs or fled farther away, to where the climate or pace of life is more acceptable. They have been replaced by a new breed of confirmed New Yorkers who come from little valley towns in California, farm communities in Michigan, and cities like Bangor, Maine.

I offer this demographic information as prelude to what follows, because what follows is about book publishing, which is a peculiarly New York institution. This is not to say that books are not published elsewhere. But New York City is to books what Detroit is to automobiles and Hollywood once was to motion pictures. Moreover, book-publishing New Yorkers are in large measure those confirmed New Yorkers from elsewhere.

On the other hand, not all suburbanites are lapsed New Yorkers. Many of them are native to their New Jersey, Westchester, Connecticut, and Long Island communities, which were once pleasant little towns. These people are equally ardent, equally confirmed non–New Yorkers. They look upon the city as Samuel Pepys, long ago in plague-ridden London, looked upon houses with crosses marked on their doors—an abomination to be endured, a source of contamination only if entered.

The attractions of New York City have almost petered out for me. It's not my Big Apple, or else I'm not much of an apple eater anymore. I remember the city when I was growing up as a much nicer place. I was born here and grew up a typical city boy, but I think I stay mostly from inertia. That and my job. I am in book publishing—New Yorkers are "in" things; in publishing, in advertising, in stocks and bonds—and book publishing, as I've said, is a New York thing.

Which brings me to Redwood Press, where I work. Some publishing houses have moved to fancy new quarters in glass-encased skyscrapers along Third Avenue, or Sixth—only a non-native New Yorker would refer to it as the Avenue of the Americas. But many houses remain on what was once "Publishers' Row," lower Fourth Avenue, or as it now says on the street signs, Park Avenue South. New York is a great place for uplifting shabby streets with fancy names, as if publishers, of all people, can't tell a book by its cover.

Redwood Press occupies the third floor of an old loft building on Park Avenue South, a few blocks north of Union Square. The offices are cold in the winter, hot and muggy in the summer, and comfortable about one-tenth of the year, on days you'd rather be outdoors anyhow. The premises were designed by the same architect who worked for Scrooge and Marley, and the furnishings were supplied by an office equipment firm that deservedly went out of business. My own office—one of the better ones, because I have a window on the avenue side—just about holds my desk, a cluttered work table, and a wall of shelves. The view from my window is depressing enough to keep me working at my desk. If I swiveled my chair around and looked out for any length of time, I'd probably be inclined to rewrite *The Shame of the Cities.*

In terms of the business, Redwood Press is what is known as a trade publisher. That means we publish books that are sold in bookstores, as opposed to textbooks or mail-order books, those picture-laden volumes, often in series, that many big magazines have gotten into. We're a small publishing house, producing about fifty books a year. I am one of four editors. The publisher, Emory Redwood, acts as his own editor-in-chief. I might add that all four of us are acquisitions editors; that is, we acquire manuscripts. We take agents to lunch and solicit manuscripts or book proposals from them. There is a direct ratio between the number of lunches we buy and the number of manuscripts we receive.

Emory has a partner, Mr. Snap, who is what is known as a "money man." This could mean that he put money into the business, the only reasonable explanation for his presence, but in our case, it also means that he handles all business, or non-editorial, matters, We also have a sales manager, who sells to the wholesalers, but no sales force. Our books are sold by independent salesmen and sales agencies, all of whom handle other small publishers, and who work on commission. We have a production manager and an advertising manager, but no director of subsidiary rights. This last is the person who makes those big deals with book clubs and paperback reprinters, deals that Emory Redwood feels he can make better himself.

Various assistants, secretaries, and clerical workers round out the staff. Actually, they outnumber the "creative" staff, because every book of three hundred pages seems to generate about three thousand pages of correspondence, memos, and billing. But the nine "creative" people, including Mr. Snap, in reluctant recognition of money's role in the creative process of book publishing, are the ones who meet in session as the "editorial board." Every publishing house pays homage to the sacred ritual of a weekly editorial meeting, and spends the other four days finding ways to circumvent its decisions.

Emory Redwood, as "head of house," chairs the meetings. On Wednesday mornings, as close to ten o'clock sharp as they can manage, the other eight arrive in the conference room. Punctuality is part of the obeisance to the sacred ritual. Besides, if Emory is not the last man in, so he can nod, smile, glower, or whatever at eight upturned faces as he strides to the head of the table, it puts him out of sorts. And Emory *in* sorts is more than enough to meet with at ten in the morning.

"Good morning, children," he greets us benignly on this day. Then he smiles broadly to confirm that he actually means the unaccustomed greeting. When Emory is being pleasant, his Viennese accent becomes more pronounced. This morning he sounds like Otto Preminger *mit schlag.*

"What goodies do you have for me today?" But the question is rhetorical; no one will open the meeting until Emory nods in his or her direction. He knows this as well as we do, and goes on.

"I myself have nothing to bring up at this time. There are, of course, projects in the works, but I cannot say anything about them at the moment."

We know all about such projects, or think we do. It has been rumored for years that Emory is being financed by the CIA. Redwood Press has featured on every list a number of books that the librarians classify as "political science." Many are by distinguished scholars or minor European diplomats; all are anti-Communist in tone.

"I will probably have to go to Washington over the weekend." He sighs, as if the burdens of work were too much. He is always going to Washington, confirmation in our minds of the CIA connection.

"Perhaps next week I can tell you more."

We can wait. There are always such projects. CIA or whatever, Emory has very good Washington connections. He arrived in this country in the late thirties, a ten-year-old Austrian refugee named Emerich Rothwald. When he became a citizen and served in the army, after World War II, he was something or other in Intelligence. That's when he must have begun developing his connections. Now he seemed to be repaying the country that took him in by publishing certain books.

Emory smiles again, looks around the table, and nods at Myra Palmer. She is a tall, thin, angular lady with white hair and big white teeth. She has been an editor for many years, the last three of them at Redwood Press.

"Nothing new." She flashes her big teeth at Emory. "That project we discussed last week . . . I've told the agent twelve-five and not a penny more. She's looking for fifteen."

"And who is the agent?" Emory asks with an edge to his voice. He has been trying to remember the project, let alone the agent who is representing the author.

"Abigail."

Emory nods, and his smile turns to a wince. Myra invariably mentions agents by their first names only, a habit Emory detests. He dislikes all vague or cryptic remarks, which I find odd in someone who was once in Intelligence.

"See that you stick at twelve-five, please." He shifts his gaze to Paul Ostrow, sitting at Myra's right.

"What has our senior member to offer?"

Paul is probably in his late fifties, which would make him the oldest person at the table—if anyone could be certain of Mr. Snap's age. But he takes the reference to himself as "senior member" unkindly. He is also the senior editor in terms of service, having started with Emory thirteen years before, at the founding of Redwood.

He starts talking about a manuscript submitted by an agent—it's a group biography of the pre-Raphaelites. There is a general discussion in which everyone at the table takes part. This is unusual and, surprisingly, leads to the decision that we will make an offer. Only if Emory is decidedly negative does everyone usually participate. Then each of us seems compelled to add another nail to the coffin of that particular project. But now the discussion is positive, and ideas flow forth on editorial, sales, advertising, and production possibilities.

"How did you ever get Clara Ford to offer us such a goodie?" Emory asks.

"Clara can be considerate," Paul replies testily.

"The only considerate thing she ever did for publishing was to poison her husband and assume control of the agency."

Having thus disposed of an agent he doesn't like, and having returned to his accustomed acid tone, Emory is pleased with himself. He enjoys his jibe and everyone's shocked murmurs, and the discussion concludes with Paul being allowed to make a decent offer to the agent.

Next it is Minnie Heffernan's turn, but the euphoria doesn't last. Minnie is a plump little pouter-pigeon of a woman with an almost inaudible voice and a host of mannerisms. Her eyes bulge, even in repose, and when she speaks, she blinks them incessantly. She also punctuates each of her short, mumbled sentences with deep swallowing sounds, so that she gulps rather than speaks. And she jerks her head to one side for emphasis. The combination of blinks, gulps, and jerks drives Emory frantic. Minnie is aware of his discomfort, but I doubt that she knows he once remarked, "Five minutes of watching that woman speak makes me break out in a sweat." If so, she would have quit. Emory's crudity appalls her and keeps her in a constant state of near-quitting.

As it is, Emory's crack about Clara Ford has offended Minnie and put her off in the presentation of her book proposals. They are not very promising book ideas, anyhow, but she does nothing to help them along. At best she provides openings for Emory's cruel witticisms. She retires from the battle with the consolation of having put him in a good mood, which will make the rest of the day easier for all of us.

It is my turn. I take a deep breath, so that my first sentence will come out loud and clear. I have found that, with Emory, what I say counts less than how I say it. If I sound assured and confident, he seems to accept what I say as worthy. At any rate, if I sound hesitant, even my best ideas are doomed.

"And what does my idea man have for me today?"

That's a joke I feel has worn thin. Some time ago I suggested that instead of picking up what the agents had to offer, it would be better to think up good book ideas and find authors to write them. Ever since, I've been Emory's "idea man." But he never negated the suggestion.

"I have a very promising book idea for Hartley Dobbs."

"And who the hell is Hartley Dobbs?" Emory snaps back, as I expected he would.

Several people around the table begin offering explanations all at once. Emory is barraged by disjointed phrases: "TV host"; "Crime Cabinet"; "hottest thing on television."

He raises a hand in protest. "Hold, hold! One at a time, please. Now, Mr. Miller"—he glares at me—"you explain. And explain it carefully, so that even I, the only ignorant member of this assemblage, will understand as well as everyone else."

So far so good, but I will have to proceed cautiously. Emory is always upset when everyone around him knows something he doesn't. But it doesn't make him angry. He genuinely wants to be "in the know" about things, particularly something all his employees know about and he doesn't. On the other hand, I'm on dangerous ground with a television personality. Emory dislikes television, insists that he never watches it and in fact rarely does, and firmly believes that television is bad for books.

"Hartley Dobbs is a long-time screenwriter," I begin, stressing a point that would appeal to Emory. "Some time ago he began writing for television, and was rather successful. Lots of critical acclaim and some awards." I see Emory's nose wrinkle at the thought of television critics, and hurry on. "This season he became the host of a new series called 'Crime Cabinet.' It's dramatizations of famous crimes, all very well done—not just recent stuff, but classic crimes as well, almost always murders. And the show has climbed to the top of the ratings."

I pause for the expected comment from Emory.

"People who watch television don't read books."

"*Some* people who watch television don't read books," I reply. "But more to the point, people who read books also watch television."

"*I* don't watch television."

"But everyone else in this room, every one of us a book reader, watches television. We all watch, and together we all read more books than you do."

"That's specious reasoning," he snorts.

"Specious?" I raise my eyebrows in a look of pure innocence.

"Well, there's something wrong with your argument. I'll put my finger on it in a moment."

"What's important," I continue, ignoring his poised finger, "is that Dobbs is a personality instantly recognizable to millions of people. And better still, from our point of view, he's a writer. A damn good writer. He writes all his own material for the show. And the audiences love it."

"If he's so good"—Emory prepares to play his trump—"why hasn't somebody snapped him up already to do a book?"

"Because," I respond softly but deliberately, a bit of dramatics on my part, "he hasn't come up with an acceptable book idea. Acceptable to him, that is."

"And you have such an idea."

"I believe I do."

He glares at me for a moment, his eyes narrowing. I stare back. Around me I am vaguely aware of seven hushed figures, but I don't want to look at them.

"You're playing with me," Emory whispers in his most Viennese tone. "I don't like it when you play games."

"The other night on his show," I say in a normal voice, "in his closing remarks, Dobbs compared the dramatized murder with the unsolved Oakes case. It was only a passing comment, but it gave me an idea. The murder of Sir Harry Oakes on his island in the Bahamas is one of the great unsolved murder mysteries of our time. Obviously it holds a fascination for everyone interested in famous murders. It also seems to have a particular fascination for Hartley Dobbs."

"You think he'll want to do a book on it?"

"I think it's just the bait to hook him."

After that the discussion becomes general again. Some people support me; some people voice the kinds of objections they think Emory would raise; some play both sides. There is nothing personal in it; office politics is, I think, just another reflection of society. Besides, when you work for someone like Emory, you're so busy watching out for him that you don't have much time for screwing your neighbor. There has always been remarkably little backstabbing at Redwood Press.

Emory puts in objections from time to time, but with waning conviction.

"Does he have an agent?" he asks at one point.

"Sidney Thorne," I reply.

"I see you've been doing your homework," he mutters. But I take that to be a compliment.

He makes one final attempt to spot some weakness in my proposal. After all, he is on record as having called television the enemy, and here he is, drifting toward publishing a book by a television personality. He turns to Mr. Snap, who sits at his immediate right.

"Have you watched this show, Fred?"

Snap, unused to being asked a purely editorial question, blinks his eyes and clears his throat. "Um, yes. Yes, I have."

"What do you think of this Dobbs fellow?"

Snap considers the question. "No snap decisions" is an office joke. He seems unaware of sixteen eyes upon him. "Talks too much," he finally responds. "Talks like a typical Englishman— like a fairy with mush in his mouth. Uses too many big words for me. Only watch the show because my wife likes it. Kids like it. I never get to watch what I want."

That seems to settle it. The discussion peters out with some specific instructions from Emory, mostly about how much to offer.

And that was how I came to meet Hartley Dobbs.

"It's an interesting idea," said Sidney Thorne.

" 'Interesting' is a noncommittal word."

"Howard, you know I can't commit my client until I've spoken with him."

We were discussing the Hartley Dobbs project over lunch. Some agents prefer restaurants near their offices, some insist upon lunching at the fashionable places in the Fifties, where they can be seen. But Sidney is willing to travel once in a while, if he knows there's a good meal at the end of the trip. And he knew from experience that the Italian restaurant two blocks from my office was worth a trip downtown.

"Listen, Sidney," I cautioned, "all I ask is that you present it to him the way I presented it to you."

"I wouldn't do otherwise."

"Well, you don't sound convinced."

"You know me, my friend." Sidney's soft, relaxed tone took on a touch of the Southern country boy he claimed to be. "I don't push my enthusiasms the way you frantic city people do."

"Sidney, you've lived in this city for God knows how many years. You're ten times more cosmopolitan and sophisticated than I'll ever be."

"True," he murmured. "But I will not partake in this city's gross mannerisms, especially excessive displays of emotion."

"I know, I know. It's taken me five years of dealing with you to realize that you display enthusiasm by not falling asleep."

He looked away and appeared to address the ceiling. "Nor do I engage in cutting remarks."

"Sidney, would it be better if I spoke with Dobbs directly?"

His gaze returned from the ceiling, brushed past me, and fixed on the dessert wagon near the far wall.

"Are you sure," he asked, "that Emory will spring for that amount of money? It's just like him to balk at the last minute."

That was quite true, but Emory had given me a negotiating

range and I hadn't exceeded it. In fact, I had kept within $2,500 of his top figure, just to show him that I was a tough bargainer.

"We discussed it at an editorial meeting," I said. "I have seven other witnesses."

"Witnesses!" Sidney actually laughed. "When did that ever deter Emory? But I'll take your word for it. Dobbs is doing very nicely with his television deal, and it will take time away from his principal—and lucrative—source of income to write a book. It has to be worth his while."

"We've made a very decent offer." I was beginning to wonder if I should throw in the other $2,500. But no. It was a small fraction of the total offer, meaningless at this point, and I wanted to bring in the book for something, *anything*, under Emory's top figure.

"I'll speak with Dobbs," Sidney said, tossing his napkin on the table. "If he likes the idea, I'll have him get in touch with you directly."

I signaled for the check. There was nothing more to discuss, and this was not a scouting lunch, where we would go over various book proposals and authors who were looking for projects. I had called Sidney about a specific matter; we had talked it through over a satisfactory lunch. Some other time we would chat pleasantly, and perhaps aimlessly, for an hour or two.

As we walked back toward the office, enjoying the bright spring afternoon, Sidney was reminded of something.

"This is Friday, isn't it?"

I nodded. It had been a lucky break to be able to lunch with Sidney on such short notice, only two days after I called him.

"Is Emory in the office?"

"I don't know," I replied. "He's supposed to be leaving early today." I really didn't want Sidney and Emory chatting together at this point. My expression must have revealed what was on my mind, because Sidney put his hand on my shoulder.

"Don't worry, Howard. I have no intention of coming up. I would never follow an Italian meal with Viennese pastry. I was just wondering if Emory was in Washington."

"He's supposed to be flying down to Washington this weekend. Why do you ask?"

"Just speculation. I heard an interesting bit of information about that Russian scientist, Kuzatov. I wondered if our Viennese superspy was down with his CIA bosses."

"Emory's not a spy," I argued. "Besides, no one's ever proved any CIA connection with what we publish."

"You keep your fantasies"—Sidney waved at a passing cab—"and I'll keep mine."

We shook hands and he thanked me for lunch as he climbed into the cab. I watched it pull away, Sidney waving a cheerful goodbye through the window. He had left me perplexed and was obviously enjoying it. I stood for a few minutes on the busy streetcorner, only vaguely aware of the traffic in the street, of the pedestrians rushing past me, of the gorgeous afternoon.

Once I'd gone upstairs, my perplexity, spurred by curiosity, turned to restlessness. Nothing seemed right. For one thing, it was precisely the kind of day when I preferred to be outdoors. For another, I had deliberately walked past Emory's office on the way to my own, to see if he was in. He wasn't. Before settling down to some desk work, I opened my windows for some fresh air and a gust of wind blew the papers off my desk. I picked them up and dropped one, which wafted out the doorway and into the hall. At times like this—and on numerous other occasions—I wished that we had doors to our cubicles.

I was retrieving the elusive piece of paper when Milt Foster came down the hall. Milt was our sales manager, a booming, cheerful, talkative little guy, just the kind of ebullient person I needed least at that moment.

"There you are," he boomed. "I was just coming to see you. What's the matter? Office so crowded you have to work in the hall?"

Ignoring Milt, which in no way offended him, I returned to my desk. He followed me into the office and plopped himself down in the lumpy chair that faced my desk.

"God, this chair is uncomfortable."

"That's to discourage visitors."

"You're not getting rid of me until I tell you how wonderful that idea of yours is, that Hartley Dobbs proposal."

I refrained from asking him where his enthusiasm had been

on Wednesday morning. Milt had been more or less on Emory's side of the discussion, but then a sales manager is generally cautious about new book ideas. That never stops him from talking the book up, once it gets on the list.

"We can use a big personality book. It's something Emory has always stayed away from, and frankly, I think he's wrong."

"I wasn't thinking of it as a personality book," I replied. Actually, I was on Emory's side of that argument.

"I know, I know. You see it as a crime book. But Dobbs is the hottest thing in TV right now, and you can't sneeze at that. TV sells books, despite what Emory says. And besides, true crime is hard to sell."

"What about—"

He waved an arm impatiently. "Yeah, yeah, I know. *In Cold Blood, The Boston Strangler, Helter Skelter.* That's three. You know how many true-crime books didn't make it?"

"So?"

"So! You know how many personality books *do* make it? It's one category where you can hardly miss. The more I think about this one, the more I'm sure we have a winner."

"We haven't got it yet." I seemed determined to dampen his enthusiasm.

"What did the agent say?"

"He's interested. He'll speak with Dobbs."

"That means he likes it. You'll see. Sidney Thorne knows a good property when it jumps into his hot little hands."

The hot little hands seemed to me to be Milt's. Why couldn't I play up to his enthusiasm? It's always good to have the sales manager backing one of your books.

"Milt," I changed the subject. "You know everything that's going on in publishing."

"So everyone says."

"Well, you say it yourself often enough so I'm coming to believe it."

"You'd better believe it. What's on your mind?"

"Something came up today. Maybe you can confirm it." I knew why I was edgy; perhaps Milt could be helpful after all. "What do you know about Kuzatov?"

"The Russian scientist fellow? The dissident?"

I nodded.

"Only what I read in the papers. You probably know as much about him as I do."

"I don't read the papers as carefully as I should. Is there something special, something recent?"

"Like what?" Milt was upright in the chair now, his antennae up and waving. I appeared to be the one with the information.

"Like a manuscript." That was only a guess on my part, but the reasonable one in light of Sidney's remarks.

Milt's face puckered, wrinkling his eyes. But I could see them darting about, as if he were looking for something in the vast file cabinets stored inside his head. He put his chubby, flat-fingered hand across his mouth and chin and massaged his face. Then he gazed hard at me for a moment.

"And Emory left for Washington."

"You're reading my very thoughts," I said.

He jumped up and headed for the doorway. "I'll get right on it. 'Foster never fails' is our motto."

After that my mood lightened. The afternoon was half over and the weekend lay ahead. It promised to be a beautiful one. And it was silly to be annoyed about someone outside the office knowing more about what was going on than I did. It happened all the time, even when it involved book projects at Redwood Press. It was typical of Emory to keep his own people in the dark while engaging in activities that he couldn't keep secret from others. I worked steadily at cleaning up the pile on my desk, and by five o'clock had done enough to make me feel as if I were caught up. As a rule, nobody but Emory would think of staying beyond five o'clock on a Friday, particularly a beautiful spring Friday, so I packed it in for the day.

The thought of Emory nudged at my buried moodiness. I wandered over to his office, to see if his secretary, Fran Bishop, was still at her desk. She wasn't, and her neatly cleared desk indicated that she was gone for the weekend. Fran was the only acknowledged secretary; the rest of us relied on typists from the bullpen and answered our own phones.

I knew that what I wanted to do was saunter into Emory's office and pick up some meaningful clue—like a notation on his desk calendar for this afternoon saying "Kuzatov," or better yet, "CIA." But one doesn't saunter into Emory's office in the master's absence. Fran Bishop, ferocious guardian of the *sanctum sanctorum*, made that abundantly clear on every possible occasion. Her presence, even in her absence, was enough to deter one.

I hovered in the doorway and settled for wandering some more. A mellow silence enveloped the offices. Emory was fond of saying that a busy office is a quiet office. But that was a different kind of quiet, filled with the clacking of typewriters, the scratching of pens, and the rustle of manuscript pages being turned thoughtfully. Emory would have been great on a Southern plantation, extolling the murmur of field hands picking cotton, or on a galley, delighting in the dip of the oars and the crack of the whip. But this was the unproductive silence of offices that were truly empty.

We occupied the entire third floor of our corner loft building. Emory had the corner office, with two windows on the avenue side—the side with the nicest view—and two on the adjacent wall, facing the narrow street. His was the only office with walls that went right up to the ceiling, and the only one with a door. This was because it was the only office with an air conditioner. Those of us with outside offices had fans to circulate the humid air, though Fred Snap eschewed such sybaritic contrivances. If it was warm, he opened his window; if it was hot, he removed his jacket; and if it was very hot, he wore a short-sleeved cotton shirt that looked suspiciously like a pajama top. Fred Snap in his summer shirt was enough to make Emory apoplectic, enough to keep him out of Snap's office, which may have suited Snap very nicely.

Six-foot-high partitions had been erected between the windows, creating the semblance of offices. Five of these cubicles ran along the outer wall on the avenue side, culminating in Emory's walled sanctuary, and four more continued down from Emory's corner along the building's street side. Each of us had an office roughly eight by twelve feet, with walls reaching less than halfway to the ceiling and a doorway partition, without a door. Two walls of floor-to-ceiling partitions created two halls onto which the offices faced. The walls screened off, visually but not audibly, the bullpen of typists, billing clerks, and other office personnel. The reception desk and entrance to the offices was at the corner of the bullpen farthest from Emory.

After hovering in Emory's doorway I wandered down the street side of the building. The offices on this side had the worst view: they faced a larger loft building and overlooked a particularly narrow and dingy thoroughfare. The first office was Paul Ostrow's. This was the homiest of all the offices, with pictures of Paul's family and signed photos of authors. I peered in briefly, wondering for the hundredth time why Emory's most senior editor was in an office on the worst side of the building.

The next office belonged to Sherwood Leitner, the advertising manager. It was one of Emory's peculiarities to arrange things so that no two editors had offices next to each other. So Sherwood, who was responsible for ads, catalogues, jackets, publicity, and promotion, separated Paul from Minnie Heffernan. Poor Sherwood was not only cramped for space, but worked in a constant noisy whirl compounded by flying visits from Emory, who screamed at him more often than not. In the late-afternoon stillness, Sherwood's untidy office seemed to be merely suspended between bouts of noisy bustle.

Minnie's determinedly feminine touch was evident even from a quick peek into her office. I avoided entering her office unless it was absolutely necessary. She had provided a hundred little touches to the eight-by-twelve cubicle that made it the epitome of Victorian maiden-ladyness, and I found the effect stifling. Emory had once said, "I know she has goddam chintz curtains on the windows that she takes down just before I come in." For once I was in sympathy with one of his cutting remarks.

The fourth office along the street side was as distinctively different as the others. It belonged to Roger McGraw, the production manager. Quick but unhurried, precise but almost fluid in his movements, Roger was one of those thin, dark, quiet people who somehow always strike me as incipient time bombs. Just to watch his somber, unruffled efficiency was enough to unnerve me with visions of Roger amok. The more efficient he proved to be, the more convinced I became that he was minutes away from frothing at the mouth and murdering us all. Roger's office was antiseptically neat, of course, as it was even when he was in it hard at work. I had a fancy that if we all had Muzak piped into our offices, in Roger's office the melodies would come out sounding like Bach fugues.

I retraced my steps, pausing again at Emory's open door. Then I walked down the hall and stuck my head into Mr. Snap's office. It was not particularly messy, but had an appearance of being unkempt. Perhaps it was an association I made with his habit of eating lunch in his office and making his own coffee on the hot plate prominently displayed on top of his low bookcase. He always used the same coffee grounds for two days. He stored the drained but soggy grounds overnight in a plastic bag and dumped them into a fresh filter paper the next morning.

It occurred to me that Fridays must have created a problem for Mr. Snap. Did he keep Friday's grounds in a plastic bag for Monday morning? No, I had seen him make fresh coffee on Monday. Did he take them home for Saturday? Or did he profligately dump them in an end-of-the-week frenzy? I was tempted to look in his wastebasket, but restrained myself. Instead, I proceeded down the hall to Myra Palmer's office. Myra stamped little of herself upon her daily surroundings. Hers was the least personalized office, and consequently the most empty when unoccupied.

Finally, I checked Milt Foster's office, a jumble of books, jackets, cartons, posters, and computer printouts. The last thing he had said to me that day was "Foster never fails." He seemed as intrigued as I was by Emory's business in Washington. Well, whether or not Foster failed, Miller was about to meddle. I headed back to Emory's office.

No alarms went off when I crossed the threshold, no flashbulbs popped. I looked up and around carefully, making faces at all the spots likely to conceal a camera, added a few obscene gestures, and walked quickly to the desk. Emory's calendar was buried under a pile of papers, but I doubted that it had been deliberately hidden. For one thing, it was open to the current day. There was only one entry: "6:30 Harrington."

Suddenly, I felt silly. All that sneaking around for one meaningless and probably innocuous diary entry. No "Kuzatov," no "CIA." Just "6:30 Harrington." I pushed the book back under the pile of papers, waved to the cameras, and hurried out. Back in my own office, I quickly put on my jacket and sheepishly headed for the elevator.

2

On Monday, New York began treating its denizens to another of its unheralded meteorological aberrations. Here it was the middle of May, and by ten in the morning the temperature and humidity were more like July. The radio had half-prepared me while I was shaving. "Unseasonably mild," the announcer had stated euphemistically, "high in the mid-seventies." I listened to an all-news station every morning, priming myself to face the day while showering, shaving, dressing, and stumbling around the apartment. It was one of those stations with its own weather bureau—no more accurate than any other, but a lot more picturesque in evading accurate forecasts. "Chance of showers" covered anything from a few clouds in the sky to the onset of a Noachian Deluge. So "unseasonably mild" could have presaged a warmly pleasant day. As it turned out, we were in for an unexpected and unwelcome heat wave.

I arrived at my office, after an unpleasant subway ride, at ten minutes past nine. The arrival time was judiciously calculated—just late enough to demonstrate that I wasn't a clock-puncher, and not too late to elicit a comment from Emory. Most of the others were already at their desks, though not necessarily working. I was at my desk, opening my mail, when Emory came prowling down the hall.

He stuck his head in the open doorway. "Good morning," he chirped, his accent pronounced and his smile a shade overdone.

It was what I called his "Viennese dentist greeting," though he meant it to be pleasant.

"Good morning, Emory."

"Did you have a good lunch with our friend Sidney?"

That was Emory, all business. One "good morning" was as much of the normal amenities as I could expect for that day. He didn't even enter my office, but hovered in the hall.

"He's going to speak with Dobbs. But I think he responded favorably to the proposal."

"By 'favorably,' you mean he managed to stay awake through the whole conversation."

"Precisely."

Emory was about to say something else, about money no doubt, when his gaze focused down the hall. "Good morning, Mr. Leitner," he called out, the Viennese dentist very much in evidence. Poor Sherwood, a born masochist, always arrived just precisely too late and, with a masochist's innate sense of bad timing, invariably ran into Emory.

"Good morning, Emory." I could hear the resignation under Sherwood's attempted cheerfulness. "Did you have a good weekend?"

Leave it to Sherwood to compound his errors. The question was innocent enough and probably asked without thinking. But Emory had been in Washington for the weekend, and we all knew that it was bad form to bring up that business before Emory raised the subject himself.

Emory stiffened, then turned and marched back to his office.

Sherwood came drooping past my doorway, "Oh, God," he moaned, "how do I manage to do these things?" As he continued down the hall, I could hear him murmur, "And it's only Monday morning."

Another week at Redwood Press was beginning in style.

Actually, the heat began to cramp our style right away. I had arrived somewhat wilted from my subway ride, and my pitifully small office fan did nothing to revive me. The ten o'clock news informed me that the temperature had already climbed to seventy degrees. That was Fahrenheit, but it didn't sound any better as Celsius. I keep a small radio on my desk, tuned most of the day to WNCN, an FM "good music" station. Emory allowed me this indulgence partly because the radio was playing Mahler the first time he discovered it, but also because he occasionally came in to listen to something on the hourly news broadcasts.

By eleven the temperature was pushing eighty. I was thinking of switching to my all-news station to see how their weather maven was coping with this unseasonable mildness when Milt Foster ambled in.

"Some heat," he said, slumping into my lumpy chair. "You don't suppose we could fix up some kind of conference in Emory's office, where it's nice and cool?"

"No, and I don't suppose Fred Snap would spring for air-conditioning the rest of the office."

"Fred Snap," Milt snorted. "That cheapskate won't even spring for the two cents of electricity it takes to run his fan. He'll sit in there in a jockstrap before he turns it on."

I started to shush him, but he waved me aside. "Don't worry, he's not in there. I saw him heading for the men's room. He's probably gone to soak his shirt in cold water. Never mind him, I came in to tell you something."

"About what we were discussing Friday?"

He nodded, looking pleased with himself. "Told you, 'Foster never fails.' But this one was relatively easy." He leaned back in the chair so he could glance through the doorway toward Emory's office. Then he bent forward conspiratorily and whispered, "Six-thirty, Harrington."

"Big deal," I muttered. "So you managed to sneak a look at his engagement calendar too. I waited till everyone left. When did you get in past the dreaded Fran?"

"I have my ways," he said, smiling. "I just wanted to see if you were curious enough to do a little snooping. Actually, if you did more than take a peek, you'd know that he's seen this Harrington several times since January. I don't remember all the dates when Emory's gone down to Washington, but the notation 'Harrington' seems to me to correspond to most of those trips."

"So Harrington is in Washington," I said. "I might have figured that out, given a year or two. So what?"

Obviously, Milt had something, but was going to make me sweat for it. I was willing to play along, so I sat back and waited, trying to look more interested than I was. The sweat was no problem in my sweltering office.

"The number-two man in the CIA's Soviet Russia division," Milt said with quiet precision, "is named Myles Harrington."

He was so pleased with himself I really had to look impressed or risk offending him. "I shall nominate you for Spy of the Year," I said. "You've nailed down the long-rumored CIA connection."

"You miss the point, smartass."

I really hadn't known there was a point, beyond confirming Emory's CIA connection. Besides, I was thinking of something else: Sidney Thorne's jibe about my fantasies. Despite what everyone like Sidney was willing to believe, or knew for a fact, I had always denied in public that Emory worked for the CIA. In private I allowed myself thoughts, but in public I was loyal. Now I wondered if Sidney took me for a jerk. He probably knew well enough that Emory worked for the CIA, and just as likely had been laughing at my naiveté on Friday. I was concerned, because I didn't want to lower my standing with Sidney, an agent who was good to know and very important to me at this particular time.

"I said, 'You miss the point.' "

"Sorry, Milt," I said. "I was just thinking about something."

"You looked like you were miles away."

"Sorry. So what's the point? I guess I'm pretty dumb about all this."

"The point is that Myles Harrington is not just CIA, he's number-two man in their Soviet Russia division. If the CIA was going to bring in a Russian defector, or even just a dissident's hot manuscript, who would be in charge of the operation?"

"The Soviet Russia division?"

"You bet your sweet ass."

"So, if there is a Kuzatov manuscript—and all the smart money says there is—our Emory is going to get his hands on it."

It occurred to me, fleetingly, that all this was remarkably clear and simple because I really didn't have the faintest notion how these things worked. After all, I didn't even know what the Soviet Russia division was. But Milt, who could identify its number-two man, seemed to think everything was clear and simple.

Besides, I read too many spy thrillers. So the realization that I might be oversimplifying things out of ignorance was a fleeting thought at best.

By noon the temperature had reached ninety in Central Park, which was where I longed to be. But I was due uptown for a lunch appointment with an agent. The favored lunching places of the literary world seemed to be moving farther uptown. This one was in the Sixties, on the East Side, a dismal subway ride away. At least the restaurant would be air-conditioned.

With my mind on air conditioning, I looked in on Mr. Snap on my way out. He had opened his window and removed his jacket, but had not turned on his fan. I waved in passing, but he didn't look up from the pocket calculator he was furiously punching. He was probably calculating how many hours of not operating his fan were needed to make up for every hour that Emory operated his air conditioner.

An editor's professional life requires him to have a cast-iron stomach and some means of working off excess calories. I have neither. As a consequence, I pop Maalox tablets like jelly beans and fight to contain my belly within a constantly restrictive belt. In addition to lunches, there are cocktail parties, professional functions, sales meetings, and occasional dinners. Some editors are not martyrs to their stomachs and expanding waistlines. I don't know how they manage it, and I hate them all.

Lunching and partying is the time-honored way of keeping in touch with the agents. And agents are the principal sources of manuscripts. If you have to bring in a dozen or more books a year, and you're lucky to find one manuscript in ten worth publishing, then you have to read a hundred and fifty to two hundred manuscripts a year to maintain your average. That means you have to wine and dine enough agents enough times to keep the manuscripts flowing in.

Despite the seeming pleasure of dining for two hours in a first-class restaurant on your company's money, lunching with an agent is work. It's business, for one thing, but it's also something of a struggle. Most agents, even when they're inclined to do business with you, act as if they have just run out of authors, or their authors have run out of typewriter ribbon, or they have recently heard that your publishing house is going out of business. The friendliest agents—and some editors and agents actually are friends—minimize this atmosphere of struggle. There are agents with whom I enjoy lunching, just as there are authors I don't feel like strangling after twenty minutes. But the agent-editor lunch illustrates one of the oddities of publishing. What should be a buyer's market is a seller's market. Why else would all agents be afflicted with paralysis of the elbow when it comes time to reach for the check?

I was ruminating on this perplexity while sucking a Maalox tablet, back in my office after lunch, when the phone rang.

"Hello."

"Mr. Miller?"

"Speaking."

"This is Hartley Dobbs."

"Ah, yes," was all I managed, though I put as much enthusiasm into that "yes" as I could.

"Yes indeed," he responded, accenting the "yes" somewhat in mimicry. I suddenly sensed it was going to be a good conversation.

"As you undoubtedly infer," he continued, "I have been speaking with my agent, Sidney Thorne. He has mentioned to me your suggestion for a book."

He paused, so I stuck in another "yes," more subdued, more tentative.

"I find it an interesting suggestion. I might have said 'intriguing,' but I find, with advancing age, that my enthusiasms have diminished in general. It is interesting—interesting enough to consider, interesting enough to discuss."

"Anytime," I replied. "At your convenience, of course, here or wherever you prefer."

"I have to be in New York on Thursday. Sidney tells me you frequent a restaurant where the food is good and the management, in return for your expense-account patronage, treats you like royalty. I imagine it must be thrilling to dine with royalty and can hardly forgo the pleasure."

"You're on," I said, "Thursday at twelve-thirty." I told him where the restaurant was, decided against any discussion of business on the phone, and hung up after a breezy exchange of see-you-thens.

Sidney had called the project "interesting," and now Dobbs had repeated the cautious appraisal. Interesting, indeed. He was hooked. I think I was smirking as I cradled the phone.

The temperature remained above ninety into Monday night. I got through the rest of the searing afternoon mainly on euphoria. There's nothing like the prospect of landing a big book to buoy up the most bedraggled of editors. On Tuesday morning, even though I switched the shower to cold before stepping out, the bathroom was a steamy sweatbox. My favorite weatherman was struggling with two problems: how to put a bright face on the continuing heat wave, and how to explain his failure the day before to predict what was impending. I paid just enough attention to hear the vital statistics about temperature and humidity. It promised to be a grim day for us loft dwellers.

The day turned out to be more of a mess than I expected—something of a disaster, in fact. When I arrived in the office, Mr. Snap was already at his desk, jacketless, tieless, and in a short-sleeved shirt. His fan was not on. In contrast, Emory was immured in his office, his door firmly shut, the better to hoard his air-conditioned air. Even with my window open and the fan on full speed, my office was barely habitable. Then, late in the morning, a series of grinding noises began filtering out from behind Emory's closed door. From time to time I could hear him step out of his office to holler something at Fran Bishop. She, meanwhile, was on the phone, sounding frantic. Ordinarily she spoke so that I could not possibly hear her from my side of the partition.

Paul Ostrow came by just before noon with an explanation.

"Emory's air conditioner has gone kaflooey," he stated with some concern. I would have offered such news with ill-concealed satisfaction, but Paul willingly took on Emory's worries.

"It can't be fixed until sometime tomorrow," he went on. "Emory's going to have to work at home."

He sounded as if he were about to cry and then take up a collection for the repairs.

"My heart is breaking, Paul."

"I can see that. What I came by to tell you is that we won't be able to have our regular editorial meeting tomorrow."

"Now it's really breaking."

"There will be a quick meeting this afternoon, in the conference room at three."

Preparing for an editorial meeting on such short notice was not nearly as bad as the prospect of meeting in our conference room. Windowless, un-air-conditioned, ventilated by an inadequate exhaust fan, it was an uncomfortable room at best. That "best" rarely existed; most of us smoked during the meetings and we always kept the door closed. So I'm sure the exhaust fan was working to asphyxiate us. Some of our publishing decisions can probably be attributed to carbon monoxide poisoning.

The conference room was hot enough for baking bread, though a bit too humid for something that useful. Mr. Snap's short-sleeved shirt looked like a summer pajama top he had been sleeping in for a week. To my right, Minnie was furiously wagging an outsized Spanish fan, and even immaculate Roger, to my left, was looking wilted.

Emory arrived last, as usual, carrying a large portable fan. He was in shirtsleeves, testimony to his broken air conditioner, and held the fan in one hand, the plug from its trailing cord in the other. His gaze darted around the room at baseboard level.

"Where the hell is an outlet?"

Everyone began frantically searching the walls.

"Fred," he bellowed at Mr. Snap, "why is there no goddam outlet in this room?"

"It's behind the door," Snap intoned.

"Well, why didn't you say so before?"

"I just remembered."

Emory glared at him, then went behind the door and plugged in his fan. He placed it on the floor, conveniently behind his chair, turned it on, and made a few adjustments until its action seemed to satisfy him. He turned back to Snap with a triumphant glare and sat in his seat at the head of the table.

Instead of glancing around the table, his usual signal for the start of the meeting, he stared at Mr. Snap.

"What is that you're wearing?"

"A shirt."

"God!" Emory sighed. "Why don't you get yourself a fan?"

"I have a fan."

"Then turn it on, instead of sitting in your office in . . . in . . . in . . ." Emory waved a hand at the offending garment. "This is a publishing company, Fred, not a scene out of Dickens."

I thought that was debatable, but at least it served to end Emory's tirade. He glanced around the room, and the meeting was in session at last.

"We will go counterclockwise today," Emory announced, "in keeping with the contrary nature of the day." He nodded at me over his glasses.

I reported briefly on my lunch with Sidney Thorne and my upcoming meeting with Hartley Dobbs.

"And Sidney agreed to the money?" Emory asked.

"He didn't object to it."

"He damn well shouldn't. It's a very healthy offer."

"It's less than the top we agreed to," I reminded him.

"I am very well aware that it is below, very slightly below, the ceiling you were given." Emory smiled nastily. "Would you like a good-conduct medal, Mr. Miller?"

"Just the book," I said.

Emory's eyes narrowed for an instant. Then he decided to let me have the last word and shifted his gaze to Minnie, who was still vigorously waving her Spanish fan.

"And you, Señora? Do you have something for us? A flamenco dance, perhaps."

Minnie's bulging eyes blinked several times. Her fan fluttered to a halt. "I have been reading," she gulped, "a remarkable manuscript." Another gulp. "A *most* remarkable manuscript."

She paused, either for dramatic effect or to consider how to go on, which only gave Emory the opportunity to ask if she cared to tell us about it.

"Yes, yes." Her head jerked affirmatively. "It's a new book by Audrey Burbage." Minnie's head continued to nod. "Audrey, as you may know, is a well-established juvenile author." Another gulp. "That is, an author of juveniles."

"We don't publish juveniles," Emory snarled.

"This is not a juvenile." Minnie shifted from her involuntary nodding to shaking her head. This produced a spasmodic jerk and some ferocious eye-blinks that unnerved Emory.

"What is it, then?" he asked.

"It's a novel about a young woman, a former farm girl, who moves to the city, becomes disenchanted with city life, returns to the farm, marries a childhood friend, settles down to be a farmer's wife, and realizes that after the city life, she is also disenchanted on the farm."

"Sounds depressing," Emory murmured.

"There is an element of sadness," Minnie's head began bobbing again. "Yes, yes, there is. But . . ." She paused, and I hoped she wasn't going to shift gears again. "But the author's positive tone"—vigorous nodding—"the obvious feeling of joy in nature, and the writing itself, so beautiful, so warm, gives the book a sort of hopeful tone at the end. A make-the-best-of-it tone, not downbeat at all."

"Who would read it?"

"Novels are mostly read by women," Minnie countered, "and this is definitely a woman's novel."

From across the table, Myra came to Minnie's support. "Audrey Burbage has an enormous following."

Emory, still focused on Minnie, muttered, "Among whom? Twelve-year-olds?"

"Among librarians, among others," Myra replied.

Now Emory turned to face her. "What others?"

"The parents of all those twelve-year-olds—the ones who actually buy the books." Myra flashed her enormous teeth. "Audrey Burbage is a big seller. She has also won several prizes."

"For writing juveniles. How do we know she can write for an adult audience?"

"I've just said," Minnie chimed in with unaccustomed forcefulness, "that it's a remarkable book." She blinked several times and her eyes moistened. "I was, in fact, profoundly moved."

Emory ignored her. "Paul," he said to Ostrow, who was busily fussing with his pipe, "what do you think?"

"Audrey Burbage has a great following," Paul replied, striking a match and holding it to his pipe. "Big sales"—he nodded to Myra, at his left—"and a big reputation," he added, puffing on his pipe. "Though, of course"—with a nod to Emory—"for juveniles." He puffed some more. "It will be news just to announce that she's written an adult novel." He retreated behind a billow of hanging pipe smoke.

"Someone once remarked about a woman preacher," Emory said with a grin, "something about a dog walking on its hind legs."

"Samuel Johnson," Myra shot back. "But the remark is not appropriate in this instance."

"Thank you." Emory's smile lingered. "I'm glad my editors are informed about such subjects as eighteenth-century literature, in which I am not so well versed. But why, pray tell, is the remark not appropriate, Mrs. Palmer?"

"Because Audrey Burbage is acknowledged to be an accomplished writer of fiction."

"Children's fiction."

"Good fiction," Myra replied, her enormous teeth clenched. "She has written books for children. They are good books. Now she has written a book for adults."

"It does not follow," Emory persisted, "that it is a good book."

"Why shouldn't it be?"

"A shoemaker should stick to his last," Emory replied in a tone of finality. Then, as if to lighten the strained tone of the meeting, he added, "As my grandmother used to say."

"Nothing ventured, nothing gained," Myra chirped. "As *my* grandmother used to say."

"Your grandmother is not running my publishing house," Emory shouted.

"Let's not all lose our cool," Mr. Snap interjected. "Of course, on a day like this, no one has much cool."

Emory stared at him and suddenly found a convenient target for his unexpected frustration. "Well, if you would only turn on that goddam fan in your office!"

All this had taken us too far afield for poor Minnie.

"Emory," she practically shrieked. "Emory, if you would only read the manuscript, you'd see what I mean." Her eyes bulged more than ever. "It's a first-rate book. I worked with Audrey years ago. I know how good a writer she is. I've known all along she would someday do something as good as this, as big as this."

Her voice trailed into a gulping mumble and I missed the rest. But there was no missing the passion of her conviction.

"I have no intention of reading the manuscript," Emory stated. "I am a poor judge of women's fiction, and don't care for it, besides."

"I'll read it," Myra offered.

"No." Emory waved off the suggestion. "You're apt to be prejudiced. In fact, after this, you're bound to be prejudiced."

"Do you want me to read it?" Paul asked, before Myra could respond to Emory's rebuff.

Emory considered this for a moment and shook his head.

"I'll read it," Milt said. "After all, if we take it, I'll have to sell it. Though sometimes," he added with a grin, "salesmen do better if they don't read the books."

"No," Emory decided, "nobody will read it for now." He sighed. "Leave it in my office, Minnie. When I return later this week, as soon as my office is habitable again, I'll look at it and give you an answer."

And that was it. All that carrying on, and no decision. A few people worked up, one frustrated editor practically in tears, and no decision. As a matter of fact, the mood of the meeting was such that we all were happy to call it quits at that point. Emory made some remark about it being time for him to get back to real work, and we filed out. Mr. Snap disconnected the fan, shut the conference room door, and padded down to his office with a faraway look in his eye. When I passed his doorway a minute later, he was standing in his office, staring at his still-motionless fan.

Even with Emory out the next morning, there were lingering tensions broken by one of those odd incidents that could only happen at Redwood Press. When I arrived, already wilted to the condition of an old lettuce leaf, Mr. Snap was at his desk in another short-sleeved shirt. The temperature was close to ninety, but his fan was not on.

The ten o'clock news reported a record-breaking ninety-two for that hour of a May morning. Mr. Snap acknowledged the smashed record—and Emory's absence from the office—by changing back to his infamous summer shirt. At eleven, it was past ninety-five and headed for the hundred mark. The radio was beginning to annoy me as much as the manuscript on my desk, and I suddenly switched it off. I was immediately aware of an unexpected humming noise from the next cubicle. It wafted over the partition, to my startled ears, like the sound of a jet engine.

I ambled next door with as innocent an expression as I could muster. Standing in Mr. Snap's doorway, I could see the fan in motion. It was on top of a low cabinet, against the wall separating our offices, oscillating slowly, purring in a kind of pent-up sigh.

"What are you staring at?" Mr. Snap growled, glaring at me over his glasses.

"It's going to hit a hundred," I said lamely.

"What are you, the weather bureau?"

"Just thought I'd report it, in case you're interested."

"If I'm interested, I'll call you."

Not that he was a cheerful host under the best of circumstances, but there was a reason for his hostility. He was counting money. And he did not like people in the room when he counted money.

There he sat, in his pajama-like shirt, his spindly arms and long-fingered hands moving swiftly over the top of his desk, as he sorted bills of different denominations into piles. He was pulling the money from stacks of envelopes in front of him, arranging the cash in neat piles, and fingering each bill with orgasmic passion.

"Is that from our mail-order ad from last week?" I asked.

He didn't answer, but continued dipping into the envelopes. Behind him, the window was wide open. To his left, the fan purred and waved from side to side. I watched in fascination as his fingers effortlessly plucked, fondled, and deposited the money.

"You'd think," I remarked, "more people would send checks instead of cash."

"You'd think," he replied, "more editors would be at their desks, working on manuscripts."

While I was thinking of a face-saving remark to cover my exit, disaster struck. A ten-dollar bill rose from one of the piles on

Snap's desk. It rose like a bird on the morning air, fluttered past his ineffective lunge, dodged his frantically grabbing fingers, sailed past his sweating face on perverse eddies, and floated teasingly out the window.

Mr. Snap, who looked like an ambulatory windmill, was out of his chair and lunging at the window. I dashed into the room to restrain him, or possibly (the thought crossed my mind) to hold his feet while he dangled out the window.

"Oh, my God!" he uttered in a kind of gargle. "A ten-dollar bill!"

"Watch yourself!" I shouted, pulling up next to him at the window. He ignored me, his eyes fixed on the money, which was dancing on currents of humid air, descending slowly to the street, three floors below. Just at that moment, one of the women from the bullpen emerged from the front entrance, probably off to an early lunch.

"There's Lillian," he gasped, waving frantically. "Lillian, Lillian!" he began shouting.

I grabbed hold of him, but he shook me off.

"Lillian, Lillian!" he bellowed. "Out the window! Ten dollars!"

Down on the street, Lillian paused at the sound of her name and looked around her.

"Out the window!" Snap screamed frantically, trying to get her to look up. "Lillian! Out the window!"

Behind us, from the bullpen on the other side of the hallway, I could hear a chair being sharply pushed back.

"Lillian's jumped out the window!" a woman's voice shrieked.

Suddenly there were more chair sounds and dashing feet. Cries of "Lillian's jumped out the window!" came rushing closer.

Mr. Snap was oblivious to this excitement. He was halfway out the window himself, gesturing madly, waving the errant ten-dollar bill into Lillian's hesitant hands. She grabbed it, looked up at us, and smiled uncertainly just as the rest of the office staff came charging into Mr. Snap's office.

He had turned away, sighing with relief, when he suddenly became aware of a swarming mass of people, hurtling around his desk, coming at him. Instantly, as they crushed past him, he flung himself on his desk, stretching his bare arms over the piles of envelopes, checks, and currency. For a moment he lay there breathlessly, spread-eagled in his rumpled short-sleeved shirt, snatching at his precious piles, then he began bellowing.

"Out! Out! Out! Get out of my office, dammit!"

At the window, I began reassuring people who were disappointed not to find a body plastered to the sidewalk, and explained what had happened.

"Out! Out!" Mr. Snap was screaming from his desktop.

It took another minute or two, but everyone trooped out. Glaring at Lillian, who stood in the doorway waving the ten-dollar bill, they marched past her. Poor Lillian, an elderly lady with a perpetually innocent expression, was perplexed by their annoyance. Nobody had bothered to explain to her what had happened.

Mr. Snap snatched the offending bill from her hand, muttered "Good job," and ignored her. As I escorted her from the office, Mr. Snap was already absorbed in rearranging his piles. I walked Lillian back to the bullpen and told her what had happened, managing to keep a straight face through the recitation. When I returned to my office, passing Mr. Snap's, I noticed that his fan was turned off. And that is how it remained for all that long summer.

3 Emory returned to the office on Thursday, and calm returned with him. A thunderstorm late Wednesday night had broken the heat wave, so the office was relatively comfortable. And whatever resentments were harbored over Tuesday's editorial meeting were being kept hidden. For my part, the morning was simply a necessary interval before lunch.

On my way to the restaurant I reviewed what little I knew about Hartley Dobbs. He was an Englishman born and bred—or born and carefully cultivated—and an American by choice. He had arrived in California as a fledgling writer fresh from World War II service—unspecified, but apparently colorful. In the early 1970s, when the appeal of British scriptwriters had long since worn thin for Hollywood, he had moved east. According to a magazine article that hinted at his wartime adventures, scriptwriting had grown boring for him; besides, he felt a need to return to civilization. If the writer of the article had smelled the aroma of sour grapes, he didn't mention it.

According to this same article, civilization for Dobbs somehow took the form of Kinderkamack, a pleasant New Jersey community that had more or less resisted suburbanization. It was close enough to the Big City to be convenient, and far enough away to remain a small town. Dobbs had been interviewed on the back lawn of his comfortably sized parcel of land, big enough, apparently, to include a number of large trees. There was some mention of a trim Cape Cod house, more than suitable to his bachelor needs and modest entertaining. From Kinderkamack he could commute to New York with ease, which he did frequently, or stay out of the city, which he did more often. And from Kinderkamack he had dispatched the scripts that grew into his successful television series, "Crime Cabinet." The article made him out to be not only successful, but secure and smug.

Obviously, there was more to him. After all, he had managed the leap from scriptwriter to television personality. I knew him from the electronic screen as urbane, witty, a bit acid, and professionally British. He was gray-haired and balding, long-faced and not really handsome, but ever so polished—something like an Alastair Sim trying to be Ronald Colman. What kind of person he was really, I had yet to discover.

He was already at the table when I arrived, and rose to shake hands. He was taller than I'd expected, at least six feet.

"I arrived early, the better to see how royally you are treated. I detect no sound of trumpets"—he smiled—"no red carpet."

"Subtlety," I replied. "More tasteful, you understand."

My favorite waiter, Franco, arrived. He bent forward, something between a nod and a bow, and murmured, "Mr. Miller, good afternoon."

I acknowledged the greeting and asked Dobbs if he would like a drink. He ordered a martini, extra dry, straight up. Franco looked at me and inclined his head a fraction. I nodded back and he left.

Dobbs watched this expressionlessly, his arms folded across his chest. Then he proceeded to glance all around the room, still largely empty, pausing at each large Alitalia poster that lined the walls. In a few minutes, Franco returned with our drinks, placing them carefully in front of us. He was performing beautifully.

Dobbs grinned. "Very impressive. I suspect," he said, "that your friend Franco enjoys the act even more than you do."

"He may be putting it on a bit because he recognizes you." I raised my glass. "Cheers."

"Cheers," Dobbs responded and sipped his drink. He nodded in approval. "So far, so good. And what is that concoction Franco so dramatically materialized with?"

"Brandy and soda."

"Obviously it is your usual. But why?"

"I read too many British mysteries."

He seemed to like that answer, and I detected some relaxation in his manner. Because I preferred to keep him relaxed, I decided not to talk business for a while.

Author lunches usually occur later in a working relationship, when an editor and an author have done a book together, or are at least at work on one. Here, at this lunch, we were feeling each other out, sort of trying each other on for size. When I suddenly realized this, I abandoned my prepared line of approach. We were either going to hit it off or not. If we did, then Dobbs would do the book, and we could discuss editorial matters at some future date.

We talked about the posters on the wall and the Italian cities they advertised. I had been to Rome, Florence, and Venice, and could talk about them. Dobbs explained to me the scene in the Siena poster, and told a funny story about his stay there. We ordered another round of drinks and discussed the menu with Franco, who responded well to Dobbs's expertise about Italian regional cooking.

While eating, we traded opinions about books and movies, and by tacit agreement avoided television. Dobbs told some stories about the old days in Hollywood. Over our espressos, we wandered into politics and I found myself talking with great heat about Emory's CIA connection, or my belated admission of that connection's actuality.

"Which upsets you more," Dobbs asked, "Redwood's activities or your previous refusal to admit them?"

"I don't know," I replied, genuinely perplexed. "But I certainly seem to be upset, don't I?"

"You certainly do. I'm surprised. My notion of New York editorial types—based entirely on passing encounters at cocktail parties—precluded this kind of reaction. I had assumed that all book editors were sophisticated to the point of insensibility. No revelation, however heinous, ruffles their all-knowing manner. Are you sure you're an editor?"

"That's how I'm listed."

"From New York?"

"Born and bred."

"It must be some deficiency in your genes. Getting upset, refusing to be in the know. You could be drummed out of the corps for a serious breach of sophistication, if word of this got out. However, my lips are sealed. I wouldn't want it known that my editor lacks sophistication."

I picked up my cup with what I hoped was studied calm.

"Then you will—"

He hurriedly raised his hands. "Yes, yes," he said. "But let's not spoil a delightful lunch with business. Mind you, I'm not adverse to mixing business with pleasure on some occasions. But this is not one of them. I've enjoyed the meal and the conversation, and even Franco's ministrations." He smiled at some private thought. "Italians would never make it as sophisticated New York editors; their suaveness is too transparent. Italianately transparent. It's one of the things I love about Italy. Everything that's put on is transparently put on; we and they both know it's a show and enjoy the show together."

"You may be right about the Italians," I replied, "but aren't you being unfair to us New York editorial types?"

"Suave, suave," he murmured, "*überall* suave."

"Well, at least I'm sophisticated enough to recognize a play on *Meistersinger—'Wahn! Wahn! Überall Wahn!'*"

"Don't show off, young man," he cautioned. "Besides, there is a profound difference between erudition and sophistication. Your catching my paraphrase of Wagner is erudition, but blurting it out is definitely not sophisticated."

I was too happy to feel rebuked. We had just launched into a discussion of Wagner operas when Franco materialized with two large brandies.

"Compliments of the house, gentlemen." He nodded toward the kitchen door, where the proprietor stood beaming in his white apron.

Dobbs turned and raised his glass in salute. Then, looking at me, he repeated the gesture. "Knowing enough to be a steady customer at a good restaurant is sophisticated."

The lunch had gone very well indeed.

We agreed to discuss the book project by phone and letter, and Dobbs said he would prepare an outline for my consideration. I saw him to a cab and walked back to the office, ecstatic, euphoric, and more pleased with myself than Scrooge after his conversion. I decided to treat myself to a Mets game that evening.

Of course, there was still the rest of the afternoon to get through. A normal afternoon at Redwood Press was antidote enough for galloping euphoria, but I wanted to talk with someone before the spell wore off. Milt was not in his office, Myra was on the phone, Mr. Snap was counting money, and Emory's door was closed. I wandered into Paul's office.

"You," he said, glancing over the tops of his half-glasses, "are looking too self-satisfied for your own good."

"I've just had lunch with Hartley Dobbs."

"I take it," he remarked with a smile, "we have a book."

I smiled back and nodded.

"Wonderful. I'm happy for you and for us." He leaned back in his chair and began rubbing his chin. "You know, I have enormous respect for Emory's publishing know-how—his intuition, really. But he does have his blind spots—very few, mind you. It was interesting to watch him when you brought up the Dobbs proposal. His inclination was to say no—not because of the television connection, because he doesn't like personality books. But he weighed everyone's enthusiasm, or, everyone's instant response, and I could see him allowing himself to be convinced."

"Paul," I interrupted, "why do you say 'personality book'?"

"Dobbs is a personality."

"But the book is about a crime, and Dobbs is a crime expert, and I don't see why you . . ."

Paul waited, smiling.

"Oh, hell," I muttered. "Milt said the same thing."

"Howie, you know as well as I do that if it were just a true-crime book, you wouldn't have offered a tenth of what you did." He took off his glasses and wiggled them at me. "You were fishing for Dobbs with the Harry Oakes murder as bait. But the book will sell because Dobbs is a big personality now, not because of the Oakes murder. You know that; I know that; Milt knows that; and Emory knows that, despite his objection to personality books."

"Then why—?"

"Why did Emory overcome one of his pet prejudices? His publishing instinct, let's call it. Despite Emory's tastes—admittedly a major consideration at Redwood Press—he does like to publish best-sellers, and every house can use one."

Paul put his glasses down on the desk and stared musingly into space. "You know, I think your bait hooked two fish. By thinking of it as a crime book, Emory can convince himself he's not publishing one of those dreadful personality books."

"You really understand him, don't you?"

"No one understands Emory," Paul snorted. "That includes Emory. He's one of the two best publishers I ever worked for. And they both were born into the business. Emory is an instinctive publisher. It's in his blood, and that transcends taste, prejudices, even idiosyncrasies. He knows what to do even when he doesn't know what to do—if that makes any sense."

I nodded. My first publishing job had been with such a person. He left much to be desired as a human being, but I had learned a great deal from him about publishing books.

"Paul, if I may change the subject—"

"Go ahead," he said. "I was getting maudlin, anyhow."

"Are we involved in some kind of CIA deal with a manuscript by that Soviet scientist, Kuzatov?"

It was an abrupt shift, to be sure, but it certainly froze Paul. He stared at his desk for a while, then began slowly tapping his fingertips on his glasses lying there.

"You know," I added lamely, "the dissident."

"I know." He continued tapping, which made me feel, unreasonably, that I had committed some enormous gaffe.

"I keep hearing things," I mumbled, "around town."

"I can imagine." More tapping. Then he looked at me with what was, for him, an arch expression. "I think," he said, "you are about to do me an enormous favor."

"I don't understand."

"That's nothing special," he replied, getting up and heading for the doorway. His smile seemed for himself, not me, and practically unnerved me. Why was I so upset? If I had asked a question out of turn, or trod on super-secret ground, he could just tell me to mind my own business, or ignore me. It had been done before, lots of times, even by the usually pleasant Paul Ostrow.

He headed out of his office toward Emory's closed door. As he passed me, still smiling, he poked my shoulder. "Don't make any vacation plans."

Next time, I told myself, I'll think twice before asking what I consider a brilliant question.

The rest of that Thursday afternoon was typically anticlimactic. Paul disappeared into Emory's office, where he remained for about an hour. Neither he nor Emory came in to see me. I remained jumpy. Paul's reaction had unnerved me. At one point, I felt like standing on my desk and shouting "Kuzatov," to see if the window would crash in or the ceiling would collapse on me. I didn't, and the day wore to a boring close.

I went out to Shea by subway and saw the Mets pull their usual act of blowing a five-run lead to lose by one run in extra innings. The familiarity of it soothed me—lulled me, in fact. I had quite forgotten my Thursday-afternoon uneasiness when Emory popped his head in at me the next morning.

"Come into my office, please," he said, with no trace of the Viennese dentist.

The events of the previous afternoon, including my jumpiness, came back with a rush. As I walked the short distance to Emory's office, I braced myself for a lecture. Sticking your nose into Emory's business—and that included company business that he hadn't discussed at an editorial meeting—was always good for a scathing lecture.

He was sitting behind his desk as I entered.

"Close the door."

That could be ominous. Or it could just be Emory hoarding his air conditioning. I closed the door and sat down across from him.

"Don't look so grim," Emory said. "I am not going to eat you."

He pressed his fingertips together in a kind of steeple and began flexing his fingers, bringing his palms together and apart rhythmically.

"One of the things I like least about publishing," he announced after an interminable pause, "is that everyone feels impelled to mind my business for me. Apparently the interests of our country, and particularly those which are quite properly kept secret, are of no concern to the New York literary gossip-mongers. Be that as it may, I cannot control the irresponsibility of people whose horizon stops at the Hudson River, whose politics veer forever to the left, and whose highest ambition is to be the first with some hot news. I can, however, by acting swiftly and cleverly, circumvent such irresponsibility."

Having aligned himself with William Jennings Bryan, wrapped himself in the flag more securely than a Roman in his toga, and proclaimed his genius, Emory shifted gears.

"You are going to go on a vacation," he said pleasantly.

"A vacation?"

"I hope you have not made other arrangements. If you have, please cancel them."

"A vacation?"

"Why do you repeat yourself? I have already said so. The vacation—paid for entirely by me—will be a happy concomitant of a patriotic service that you will perform. To be perfectly honest, I am surprised by my own generosity. But then, I am so pleased with myself, with what I have concocted, that I am willing to be generous—just this once, of course."

"What vacation?"

"Try not to be dense. I have some relatively complicated instructions to impart to you. They require that your head be clear, even clearer than when you conduct one of your famous negotiations with an agent. You are going to do your country a service. I would say a service to Western civilization, but I suspect you would consider that hyperbole. Think of it in more practical terms. You are going to do Redwood Press a service."

"By going on vacation?"

Emory pressed a hand to his brow. "I keep telling you, don't play games. I did not hire you as court jester. This is a very serious matter."

"I'm sorry, Emory. But I am confused. I take it this has something to do with yesterday afternoon. Paul did say—"

"Never mind what Paul said. I am doing the saying. Right?"

"Right."

"Yesterday afternoon you confirmed what I have been telling those . . . certain people in Washington. New York publishing is an incestuous business in which gossip takes the place of sex.

Everyone knows everyone else and they all talk to each other all the time, particularly about other people's business. You can't keep a secret in New York publishing."

"You mean the Kuzatov manuscript?"

"I mean any secret," he snapped. "Will you please let me continue."

"Sorry."

"Yes, Kuzatov. What is it that the gossip-mongers are saying about me and Vasili Kuzatov, or rather about Kuzatov, me, and my CIA masters?" He sneered and waved away any answer. "Yes, I know all about what the wagging tongues say—that Emory Redwood has sold his soul to the CIA. And is that what they say about Kuzatov? Kuzatov, who has a hundred times more soul than any of them."

There was nothing for me to say, so I sat and waited for Emory to simmer down. He seemed to be debating something in his mind.

"I think," he said, "I will tell you only as much as you need to know to carry out the plan that I have conceived. That will make it easier for you to avoid any unnecessary explanations. Not that I anticipate any problems, but what you don't know, you can't answer questions about."

Typical Emory. "Questions from whom?" I asked.

"Never mind. There will be no questions. And stop interrupting. Just listen carefully. There is indeed a Kuzatov manuscript. It is an analysis of how the Soviet hierarchy systematically denies the process of creative expression in Russia, not only among the creative people, but within their own ranks. This is not news, of course; we know what they are like. But it is in the delineation of how the hierarchy proceeds, how pervasive the system is, how cruel, that Kuzatov has scored a triumph. And he is, or was, himself in an exalted position as a leading scientist. Suffice it to say that this is a denunciation of Soviet bureaucracy and tyranny unmatched since Milovan Djilas's book, *The New Class*—and that was years ago.

"I say there is a manuscript. Actually, there are two. They are both manuscripts of the same book, one in French and one in Russian. By methods that need not concern you, both manuscripts have been, or are being, delivered out of the Soviet

Union. They will go out separately; they will arrive at separate destinations. The French manuscript is to be put in the hands of the I. Pierre publishing house in Paris, who will be bringing out an edition in French about the same time as ours. The Russian manuscript will be delivered to Istanbul—don't ask why—where we will pick it up. Where *you* will pick it up. Congratulations, you're going to Istanbul."

"Istanbul?"

"Don't interrupt. I know you have a passport; you made a big enough fuss last year about getting it." I had taken off from work one morning to clear up some red tape snafu about renewing the passport. So much for his big fuss.

"Istanbul?"

"And stop repeating everything. You can be most aggravating at times. Now this is important. We are going to make a big splash about the French manuscript. I am going to go to Paris. I will explain all this in detail in a minute. But we are not going to say anything at all about the Russian manuscript. You are going to Istanbul on vacation."

"How do I explain that to people?"

"You have always," he said through clenched teeth, "had a desire to see Istanbul. And you will see Istanbul with a broken arm if you persist in being difficult."

"It's just that I'm overwhelmed."

"Then *under*whelm yourself. And pay attention."

The plan had a certain madcap simplicity about it, as if Emory were making it up as he spoke, and couldn't be bothered with too many details. It was less complicated than something in a spy novel, but just devious enough to be more outrageous than sensible. I listened carefully and nodded in agreement from time to time, but underneath my outward acceptance, a part of me said it was all crazy. For one thing, Emory was springing it on me cold, which didn't give me much chance to evaluate its merits, even if I could. Moreover, Emory, in his usual way, colored what should have been a factual briefing by emphasizing points that pleased him and omitting whatever he felt I need not know. The result was a set of instructions that bordered on the unreal.

"We will have to move swiftly," he said, "and the timing is

made complicated by when we can get the manuscript to Istanbul."

I leaned forward to say something.

"If you repeat 'Istanbul' once more, I will break your arm here, now!"

I settled back in my chair.

"My own timing is easier. I can get to Paris on short notice, and the people at I. Pierre are already expecting the manuscript. The important thing is that we announce the book in Paris when we know you have picked up the Russian manuscript in Istanbul. This is important because, even though I. Pierre can publish the French text and we could more easily translate into English from that, we all feel it is preferable to have Kuzatov's original Russian to work from and check against. After all, even though he is reasonably fluent in French, Russian is his native language."

I was going to ask who "we all" were, but decided against it.

"When you get to Istanbul, you will stay at the Hilton; everyone stays at the Hilton. We will arrange for your reservation, and you will stay four nights so that you really look like a tourist. We will even book you into a good room, on the Bosporus side." He smiled. "You see how nicely we are taking care of you."

I nodded and smiled back, without knowing what the hell he was talking about.

"Of course, there is another reason why you will be staying four nights. Your arrival will be noted by, uh, the proper persons. They will need one day to make their arrangements. So you can spend your first full day in Istanbul sightseeing. I should note that you arrive in the afternoon and, what with settling in and eating dinner and getting over the effects of a very long flight, you will want to go to bed early your first night." He made it sound more like an order than a suggestion.

"So you will be fresh and eager for a day of sightseeing. Two days, really, because you won't pick up the manuscript until the evening of your second full day. Don't push yourself. I want you to be alert when you pick up the manuscript. But two days should allow you enough time to see the sights." He paused and began ticking off the instructions on his fingers. "Now remember, arrive afternoon, settle in, early to bed: night one. Full day of sightseeing: night two. Then a day of sightseeing until mid-afternoon. About six o'clock—and remember, set your watch to Istanbul time—you will proceed to your rendezvous point, a restaurant. That should please you."

Again, I leaned forward to ask a question. He interrupted his finger-counting and held up his hands.

"Patience. I will explain the rendezvous in detail in a moment. Let me continue with the itinerary. You will arrive at the restaurant between six and seven o'clock. That is early for dinner in Istanbul, so it should be relatively empty. That will make it easier to conduct your business. After dinner, return to your hotel and leave the manuscript in your room. The rest of the evening is yours to enjoy the nightlife of Istanbul." He made it sound like a line from a travel brochure.

"Night three," he went on, ticking off a finger. "Now it is necessary that you spend the next day out of Istanbul. There is a ferry trip up the Bosporus. You will enjoy the sights. There is a castle, I believe, or a fort, or something equally interesting. If you come back early in the afternoon, take the ferry across to the Asian side. That, too, is very interesting. Just make sure you do not return to your hotel until dinnertime. You should be tired out by all this wandering around, so I suggest an early evening." He jabbed at a finger. "Night four. The next morning you can head out to the airport and fly home."

We stared at each other for a moment. Emory seemed very pleased with himself. I must have looked blank. It had all registered, and it hadn't. I had heard the words, had even understood them. But I couldn't associate them with me. I was reminded of one of those scenes in a James Bond movie where M. explains Bond's new assignment to him. Any moment, Paul Ostrow would walk in with a briefcase for me containing gas bombs, cyanide pellets, a secret knife, and a small rocket to fly me out of Istanbul.

"Now I will explain the rendezvous," Emory said. "Listen carefully, though I will go over it again with you before you depart. Many times, if necessary. And the address is written down here." He held up a slip of paper, but did not hand it across.

"The restaurant," I mumbled.

"The restaurant," he repeated. "It is relatively easy to locate, I am assured, being right off Taksim Square. And that, I am told, is the Times Square of Istanbul."

"That should be easy to find, then."

"Don't be overconfident," he warned. There is no winning with Emory. "Overconfidence leads to carelessness, and you cannot afford to be careless. I have not spoken of it, purposely, but the stakes in this matter are very high."

"You mean there's danger." I tried to make it a casual statement.

"Danger?" He stared off at a corner. "Not really, I should say, but anything is possible. You see, I will be deflecting the danger from you by my own highly publicized trip to Paris. That's where this plan is clever. All eyes will be on me while you conduct the crucial business. It's like a magician confounding his audience by distracting their attention from his busy hands."

He leaned back in his chair, cheerfully contemplating his cleverness.

"No danger, then, really," I reminded him.

"Precisely. If you are careful and follow instructions." He handed me the slip of paper. "And I am sure you will. Here is the address of the restaurant. You will arrive between six and seven. Earlier in the day, you will have purchased some innocuous books—I suggest guidebooks—which you will have wrapped in a bag or package the size of a three-hundred-page manuscript. This package you will take with you to the restaurant. When you arrive, check the package with the proprietor. Eat your dinner, which should take less than an hour. In any event, leave by eight o'clock, when the restaurant should begin filling up. And for goodness' sake, remember to take the package with you when you leave. The manuscript will be in it, instead of your guidebooks."

"How will the . . . I mean, who . . ."

"No questions; the less you know, the better."

"But how will I be sure the transfer has been made?"

He considered this briefly. "All right, get the guidebooks put in a bag instead of a sealed package. That way you can glance into the bag when you leave the restaurant—*before* you leave the restaurant—and make sure the manuscript has been substituted."

"Just like that?"

"Certainly. Be casual when you glance at the bag. It's only natural that you would check its contents. Casually, of course."

"What if the manuscript isn't there? What if there are still just guidebooks in the bag?"

"Proceed as planned. Exactly as planned. There are contingency arrangements. I need not burden you with them. That is part of the reason why you should be out of Istanbul, far from your hotel, all through the following day."

"And if the manuscript is in the bag?"

"Take it back to the hotel. Leave it in your room. Do not secrete it in some terribly clever place. On the other hand, do not just throw it about. Do with it what you would do with a package of guidebooks or some other souvenirs. And spend the next day on the Bosporus, as planned. On the Bosporus, whether or not you have picked up the manuscript."

"You make it sound as if something were going to happen that day."

He slammed his hand on the desk. "Nothing is going to happen. You have an overactive imagination. What other people may or may not do on this end . . ." His voice trailed off. "Remember," he insisted, "all the attention is going to be on me. As far as anyone is concerned, there is only one Kuzatov manuscript—in French, in Paris, and I am going over there to pick it up. You are going to Istanbul on vacation. You will be just one more American tourist admiring the Golden Horn."

"Well, it sounds simple, the way you say it."

"It is simple. Howard, I promise you there is no danger. I was going to send Paul, a family man. Would I have endangered an old friend, a man with a wife and children?"

I considered answering that, but he looked so patently innocent, even a bit hurt, that I gave up.

"It is not a question of your being more expendable," he continued. "But rather that you are so . . . so unlikely; and besides, it is important to have Paul here to run things in my absence."

That was weak. He didn't let anyone "run things" in his absence. But I was oddly convinced by his remark about my being so unlikely. That made sense. The unlikely spy, or courier, or whatever it was called. I was unlikely, the situation was unreal, and it never even occurred to me to ask what would happen if I refused to go.

"You are doing a service to your country and to Redwood Press." He smiled broadly. "See, I am putting the country first."

I smiled back, weakly.

4 When Emory handed me a free, if unlooked-for, vacation, I was too confused to consider his offer in any detail. My confusion, I must admit, was self-induced, the result of seeing myself as a sort of James Bond in the middle of a Graham Greene situation. When I wasn't hyping myself with that fantasy, I was, more reflectively, casting myself in an Eric Ambler novel—one of those innocent travelers who gets mixed up with foreign spies and never understands what's going on around him. None of this—and I say this in pure hindsight—served to allay my confusion. In other words, I was maneuvered into the trip and didn't have the sense to know it.

Probably that was all to the good. I could occupy myself with looking at, but rejecting, fortunately, several expensive trench coats, and reading books about the Ottoman Empire and modern Turkey. I bought a guide to Turkey and pored over the pages devoted to Istanbul. And I finally purchased a pair of jogging sneakers I had been promising myself—in the vain hope they would force me into jogging—on the grounds that they made excellent, lightweight walking shoes. All the while, I could keep from paying too much attention to the more specific matter of what would happen when I actually got to Istanbul.

Then too, there was the normal course of my daily work to occupy some portion of my frazzled mind. I did manage to get through the everyday chores that—face it—make up most of a person's day. However, without meaning to digress, I must report that my thoughts over the next few weeks were not entirely occupied with my upcoming trip. Something—a couple of things, actually—grabbed some of my attention.

If my thoughts were not entirely occupied with my upcoming trip, there were two pretty big distractions to account for this. One was the ABA convention, the annual American Booksellers Association convocation and grand bash that is to book publishing what a good auto de fe was to the Spanish Inquisition. The other distraction was Dinah Foxworth.

I was at a party, a non-publishing party, given by two old friends, when my hostess grabbed hold of me and steered me toward a somewhat familiar figure. I couldn't quite place her, but I knew that I was supposed to know her. She was short and trim, rather pretty in an unexciting way, with short blonde hair billowing around her animated face. The vivacious air that she projected seemed a part of her general animation. She was speaking with someone, and her flow of words was accompanied by smiles, gestures, and an almost hopping motion.

"Excuse me, darling," our hostess interrupted, shoving me closer, "but you two should meet. That is, if you don't already know each other. You're both in the same business."

The blonde directed a ready smile in my direction, and suddenly I remembered who she was. I had seen that smile flashing up at me from the pages of *Publishers Weekly* and across various crowded rooms at publishing functions.

"Not really," I mumbled.

"Not really in the same business?" she asked.

"No, I mean not really know each other." My tongue seemed to be all thumbs. "Though, of course, I mean, that is, I know *who* you are."

Her smile evaporated. I tried a weak one of my own. Our hostess finally came to the rescue.

"Dinah Foxworth," she said airily, "Howard Miller."

With a wave of her hand she was off to some other couple. The man with whom Dinah had been talking wandered away. We stared blankly at each other. I didn't know what to say. Dinah Foxworth was the director of subsidiary rights at one of the bigger publishing houses. But more than that, on the strength of a recent string of million-dollar-plus sales to various paperbackers, she was the reigning queen of subsidiary rights. The *Times* had called her that just a week before, and I felt as if I were in the presence of a Personage.

"Then you are in publishing?" she asked.

"Yes."

"Where?"

"I'm an editor at Redwood Press."

"He's a nut," she said firmly.

"Emory?"

"And a fascist."

"Not really." I tried to think of something to say in Emory's defense.

"I was on a committee with him," she went on, "and it was perfectly awful. He was rude, overbearing, and insufferable. And if a woman, a mere *woman,* was brazen enough to voice an opinion . . ." She shook her head angrily. "All that damned German arrogance."

"He's Austrian."

"Same difference. I don't know how you can stand him."

"Oh, there are ways of managing. Can we talk about something other than Emory Redwood?"

"With pleasure."

"Can I still congratulate you on your success? Or is that old-hat?"

She closed her eyes and sighed. When she opened them, the baby-blue seemed to have darkened perceptibly. "You too?" she said. "You're in publishing, at least you should know better."

"I seem to have said the wrong thing."

"Look, we publish more than two hundred books a year. So I sell three of them for big money. There are still two hundred others. And most of them are small-ticket items. Twenty-five-hundred- and five-thousand-dollar items. It's a lot harder these days to sell the twenty-five hundreders. And there are a hell of a lot more of them."

"Don't I know it. I specialize in those."

"I wouldn't brag about it, sweetie. Editors like you, people like me can do without. Have you got something against best-sellers?"

"I have a friend," I said by way of reply, "an editor, who has a large hand-lettered sign on the wall facing his desk. It says: 'Publishers don't know what makes best-sellers; if they did, they would only publish best-sellers.' "

"That's hardly original," she countered.

"I know. Let me put it another way. If you ask an editor what's his favorite book of all the books he's worked on, I'll bet you that practically every editor will name some book you've never heard of, or just vaguely remember."

"Unfortunately, I think you're right." She turned on her smile, a bit wickedly this time. "Thanks for the small comfort to keep me warm on a cold night."

"I only say it because it's so."

The smile grew more artificial. "You're a charmer." She began to edge away and look around the room. "It must be all that constant exposure to Redwood."

"Not really."

"Do you have some kind of copyright arrangement on that?"

"On what?"

"The phrase 'not really.' "

She turned to walk away and I said, "Perhaps we can continue the conversation on some other topic."

"Not really," she said, glancing back to give me the full smile.

It wasn't the most auspicious of beginnings, to be sure. In fact, on the basis of that conversation, I figured I had had my one and only contact with the great Dinah Foxworth. We bumped into each other two or three more times in the course of the evening, smiled politely, and ambled on. The idea of another conversation with her didn't appeal to me, and I assumed she felt the same.

The party took another two hours to wind down, and when I finally left it was late even for a Saturday night. As I stepped out onto the street, the first thing I noticed was that it was raining. The second thing I noticed was Dinah Foxworth, about two feet in front of me. She was glancing up and down the street, looking for a taxi, I supposed, tossing her blonde head and engaging in that little hopping motion of hers. She turned around, saw me, and flashed the smile.

"Do you like walking in the rain?" she asked.

"Not really," I replied.

She looked at me puzzled for a moment and then burst out laughing.

"The Not Really Kid from Redwood Gulch." She put her arm through mine. "I suppose it's better than 'yup' and 'nope.'"

"Am I walking you home?" I asked.

"I certainly hope so. I don't mind walking in the rain. But not alone at this hour."

"Then lead on; the honor of Redwood Gulch is at stake."

We strolled along in what was little more than a drizzle, a minor inconvenience on a warm night.

"Honestly," she said after we had gone a block in silence, "I don't know how you can stand working for him."

"I thought we weren't going to talk about Emory Redwood."

"You're right. We'll talk about your books. Have you ever worked on any I would recognize?"

"I've just signed up an author you'd recognize."

"Who's that?"

"Hartley Dobbs."

"Never heard of him."

I stopped, let go of her arm, and stared at her. "What do you mean, you never heard of him? Don't you watch television? He's got one of the most successful shows of the season." I was waxing indignant when I saw she was laughing at me.

She hooked her arm through mine again and pulled me down the street. "Don't mind me," she apologized, "I just like getting people's goats. Besides, I owe you one from before."

"You already paid me back for that."

"Was I rude? I'm sorry. I can't resist good exit lines. And of course I know Hartley Dobbs. 'Crime Corner.'"

"'Crime *Cabinet*,'" I corrected.

"Whatever you say. I can't afford to argue with you under the circumstances. I need you for protection."

"Some protection."

"Don't knock it. Besides, you have that boy-scout air about you."

"Oh, thanks."

But she squeezed my arm with hers and walked along smiling.

There seemed to be nothing for me to say, so I just enjoyed her presence and we continued for several blocks in silence.

When we reached her house, a renovated brownstone, she pulled away from me and held out her hand.

"Thanks for the company."

"My pleasure, I assure you."

"Good night . . ." A frown suddenly crossed her face.

"What's the matter?" I asked.

"I'm terrible at remembering names."

"Howard Miller," I said, shaking her hand.

Then, with a final smile, a warm one I thought, she trotted up the steps, unlocked the door, turned to wave, and was gone.

The annual convention of the American Booksellers Association is held in a different city each year. Publishing books may be a New York business, but selling them is national. This year the convention was in Chicago.

I'd never attended, partly because it's invariably held over the Memorial Day weekend, and I hate to give up my holiday, but mostly because I'd never been invited. There was nothing personal in this; the ABA was a booksellers' convention—salesmen were essential and editors, superfluous. But in recent years, the convention has become a wheeler-dealer jamboree, so subsidiary rights people and most editors-in-chief have begun to show up in force.

Emory was going, of course, to wheel and deal. Milt Foster, as sales manager, was also going, of course. Tony Hunt, the sales representative for Redwood Press in the Midwest, who operated out of Chicago, would be helping at the booth. Another person was usually needed at the booth and it was almost always Paul Ostrow. I didn't expect to go. But I wasn't figuring on Emory.

"You will have to help us man the booth in Chicago," he announced from my doorway, five days before the convention.

"Chicago?"

"We will not go through that routine again. If you want to repeat the names of cities, please do so on your own time."

"But I have to be in Istanbul."

"That is two weeks later. It is perfectly possible to be in Chicago, and then in Istanbul two weeks later. For that matter, it is possible to be in Chicago, and then in Istanbul two *days* later. But no such demand is being made of you. I need you in Chicago."

What he needed was to keep me in sight, under his thumb. I was annoyed, but didn't see how I could refuse. Anyhow, going to the ABA was supposed to be a plum.

"When do I have to go out?"

"Saturday morning will do."

That was a relief. The setting up of the booth was done on Friday.

"Where will I stay?"

"Milt has taken a double room; you will be able to stay there. I expect that you will be needed through Tuesday. You can stay over through Tuesday night and fly home on Wednesday."

"How do I get out there?"

"You go on a wonderful big machine that flies in the air. It's called an airplane." Emory glowered. "Really Howard, this air of helplessness does nothing to raise my confidence in you."

"It's just that . . . it's so sudden."

"Your life is filled with surprises these days."

"I've never been to an ABA."

"Then you must bring your teddy bear to sleep with you. He'll be such a comfort." Emory turned back toward his office. "And you can show him all the sights in Chicago," he called back over his shoulder.

Emory being funny was enough to set my teeth on edge. But his parting crack had given me an idea. Sightseeing of a special kind was definitely on my mind.

I reached for the phone and called my friend David Cummings. The two of us had worked together some years back and had remained friends. I felt fortunate in this because David, a very commercial editor with a wide range of interests, had gone on to become editor-in-chief at a major publishing house. He was a good man to know.

"Hello, chum," he said when I got through to him.

"Hello, David. How are you?"

"Good as can be expected. What's on your mind?"

"David, are you still conducting your architectural walking tour in Chicago?"

"Indeed I am. The ranks seemed to have thinned a bit, but I'm determined to press on, even if I have to walk alone."

"Then you won't mind if I join you."

"I'd be delighted, chum. You're more my idea of a walking partner than some of those who've dropped out. But tell me, how did you manage to get to go to Chicago? Some special dispensation from the Mad Hungarian?"

"He's Austrian."

"Same difference. But really, I'm delighted. We meet at noon on Monday in the French Impressionists room at the Art Institute. I'll see you there, if not before, and wear walking shoes."

So much for that. David had been enthusiastically planning for weeks to lead a walking tour of the famous architectural sights in Chicago, and had been lining up fellow walkers. From what I had heard, more people were applauding his project than were actually agreeing to join him, but it was bound to be a select group in the end.

The next thing was to line up parties. Although I had never been to an ABA, I damned well knew what to do now that I was going. This was a little harder, particularly on such short notice, and took a number of phone calls. First of all, I had to determine which parties on which nights were the ones to go to. Then I had to see if I knew anyone who could get me in. As it turned out, I am fortunate in having many of the publishing friends I do, not only because they're nice people, but because they all came through royally. An hour later I was lined up with parties for every evening. More than that, I had an invitation to the Playboy Mansion for Monday night. That was the hottest ticket in Chicago, so to speak, and a real coup. Of course, I now owed some favors in return, but it was worth it.

There were more mundane matters, but I wasn't concerned about them—my remarks to Emory notwithstanding. Fran Bishop, as I well knew, had arranged for my flights in and out of Chicago and my floor pass to the book convention. That left a little business with Milt. I found him in his office, behind a pile of computer printouts.

He looked up and muttered, "Hello, roomie.'

"Ah, then Emory has told you."

"Probably before he informed you, if I know our exalted leader."

"Probably you're right."

Milt glared at me. "Listen, fella, the ABA is a serious business. You've never manned a booth, have you?"

"I've manned one at a teachers' convention."

"That's something. Not much, but a start." He launched into a discussion of the ABA, advice on dealing with booksellers, instructions on how to take an order for books, special information about the list, and a lot more. I took in as much as I could and waited for him to wind down.

"There's just one thing . . ." I finally managed to interject. He peered at me suspiciously.

"I need a little time off Monday afternoon—a few hours, starting just before noon. I'll be back by three. I promise."

He continued to peer at me from under his eyebrows, weighing some answer. I was expecting a barrage of invective. Instead, a slow and somewhat wicked smile began to cross his face.

"That," he said, "is one of the more unreasonable of many unreasonable requests you have made of me. Unreasonable to the nth degree. I want you to know it because it just so happens that I have an unreasonable request to make of you. And it may very well be that we can accommodate each other."

"Go on."

"Something has come up unexpectedly, but most agreeably, that involves Monday evening—or Monday night, I should say. I was beginning to wonder how I was going to manage it . . . not *would* I manage it, mind you, but *how* would I manage it. But you now present me with a possible solution."

"What do you have in mind?"

"Howie, my boy, if you can contrive to be out of our room Monday night, and stay out—all night—then I think I can manage to see my way clear to man the booth without you for three hours on Monday afternoon. Do I make myself clear?"

"Perfectly clear."

"Can you do it?"

"Consider it done."

And so I was all set for Chicago. Of course, I didn't have anyplace to stay for one of the four nights I'd be there; but I had parties to go to every night, including the choicest one in town, and I was going on the celebrated walking tour that I really didn't want to miss. Some of the time would be spent working, but I wasn't going to think about that.

If you have ever attended a large trade show and convention, then you know what the ABA is like, only it's books instead of housewares or automobiles or hi-fi equipment. Every publisher, large and small, is there. The booths range from single stalls to elaborate display areas. The smaller ones are staffed by tired-looking couples who spell each other in shifts and wait hopefully for booksellers to stop by. The larger ones are garishly lit, abundantly staffed, crowded with attention-getting displays, and constantly filled with people. The aisles are jammed with slow-moving scavengers armed with shopping bags and sucking up free samples like ambulatory vacuum cleaners. Traffic is snarled at intervals by promenading six-foot ducks advertising children's books or by some other nonsensical obstruction. Everywhere people are greeting each other effusively, and here and there some people are actually selling books and writing out orders.

Depending upon whether you are on the floor of the convention hall, at one of the panel discussions, or attending a breakfast, the ABA is Barnum and Bailey, back-to-school, or an early-morning edition of "The Tonight Show." What with the carnival atmosphere, the parties, the trysts, and sightseeing, it's a wonder that much business gets done. But everyone pronounces each convention a vast success, and the business of selling books goes on for another year.

I had a fine time in Chicago. After a hesitant start, I got into the swing of answering questions, talking up the list, and taking orders at the booth. I found time to wander around the floor, visit other booths, and chat with friends. I went to two parties on Saturday night and managed to squeeze in three on Sunday.

Milt was in charge of the booth and rarely left it. Emory showed up for a bit each day, but was mostly scouting the competition with a jaundiced eye or off the floor on some business of his own. Tony Hunt, our sales rep in the Midwest, came by to lend a hand with the order-taking, though he had to perform this service at other booths as well, for the other publishers he represented. On Monday morning he showed up a little after eleven-thirty, nodded to Milt, beamed at me, and said, "Off you go."

"I appreciate this, Tony."

"So does Milt," he replied, "which counts for more. Enjoy your walk."

"You know about this walking tour?"

"Walter Templeton was mentioning it at the booth this morning."

"He's going?"

Tony nodded. Walter Templeton was editor-in-chief at the largest house that Tony represented. That sounded pretty encouraging in the way of walking companions. Walter was someone I knew slightly, enough to say hello to. He was as bright as David Cummings and as well-rounded in his interests, but, by reputation, more of a cosmopolite. He moved in "society" circles, got himself appointed to lots of publishing committees, and was known for his superb European contacts, bringing into his house many high-class books with an international reputation.

I was at the appointed meeting place right on time. David and Walter were there, admiring a Van Gogh I had never seen before. We greeted each other and continued to examine the painting.

"Let's give these baubles a few minutes," David said. "We're waiting for Farkas. Everyone else has dropped out."

I would gladly have given the "baubles" an hour. The room was a symphony of color—Impressionism and Post-Impressionism epitomized in one large, glorious, enclosed space.

"There you are," a voice boomed out.

I turned to see Mervin Farkas bearing down on us. Mervin was another person I knew just well enough to say hello to. He was at that moment one of the "hottest" editors in the business, someone every up-and-coming editor would want to know. Recently appointed as the editor-in-chief of a major publishing house, he had moved from a much smaller house that was known for its high-quality books, and that he had singlehandedly made successful.

So there I was, part of a foursome, the other three parts of which were prominent editors-in-chief. I was almost more excited about my companions than I was about the tour itself. I wished Emory could see me.

When we finally exited onto Michigan Avenue to examine Chicago's architecture, we were in high spirits. The day was gloriously sunny, to match our mood, and a leisurely walk seemed the perfect occupation for the next two hours. David and Walter and Mervin kept up a flow of banter to which I added what I could. If one of them particularly liked something, the other two would either object or take sides, and I would find myself drawn in, for or against a balustrade or a window. We proclaimed the tour a success and the only worthwhile reason for having attended the ABA. In the taxi that we took back to McCormick Place, the vast convention center, we discussed where the next ABA should be, and settled on Angkor Wat.

I was only a few minutes late getting back to the booth, and Milt and Tony had things well in hand.

Before leaving to put in some time at another booth, Tony invited me to a party he was running that evening. He had taken over an amusement arcade, and all the games and pinball machines would be free. I had plans, of course, and wasn't going to miss my chance to join in the best party in town, but I thanked him and wrote the address on the back of one of his cards.

Fortunately, I was in place at the booth, even taking an order, when Emory breezed by to see how things were going.

"You look a little flushed," he said. "Don't get excited by all this madness. You'll get used to it." Then he was off again.

Flushed? I wondered if I was blushing guiltily for my three-hour escapade. And then I realized. On top of having enjoyed myself thoroughly, I had gotten a sunburn.

It was about nine o'clock and I was at the Hefner mansion. Even though the security and checking procedures at the door were elaborate, and even though I had been assured that my admission ticket had been hard to come by, it was apparent that half the ABA was also at the party. The only people missing were Hugh Hefner and the Bunnies. I was pushing my way through the most densely packed crowd I had ever encountered, trying to get to the bar, and nodding and smiling to people as I nudged my way forward. I had arrived ten minutes before, and already I was wondering what I was doing there. Conversation was impossible; the incredible din of the amplified music over the babble of hundreds of voices kept me from hearing what anyone was saying. I kept nodding and smiling at people I knew, either fairly well or just by sight, and wormed my way to the bar.

"Have you seen the swimming pool?" someone shouted at me.

I shook my head and inched nearer to the bar.

"You gotta see the grotto," he added.

I made a mental note and finally found myself pressed up against the enormous oak bar. Drink in hand, I began to worm my way back to the center of the room. More smiles, more nods, more shouted exchanges I could scarcely hear. Somehow I worked my way across the large space that I assume was the living room, and through a doorway at the far end. Downstairs I found the swimming pool. About a dozen people, drinks in hand, some of them with plates of food, were wandering around, staring disconsolately at the water. There was no one in the pool. Not one single, solitary Bunny, costumed or in the buff.

A salesman I knew vaguely came up and leered at me.

"You gotta see the grotto," he said.

"So I've been told."

"Some place," he announced, staring at the water.

"Yeah." I watched him walk to the edge and peer in.

"Some place," he repeated.

I wondered how sore he would be if I pushed him in.

"But you gotta see the grotto."

I turned away and headed for the doorway.

"It's down that little winding staircase."

Almost by accident—I was avoiding a large hole in the floor of the adjacent room, a hole through which a fireman's pole descended—I found the little winding staircase. I cautiously made my way down the narrow steps into what was obviously the grotto.

It was a dark room with stucco walls. The wall on my left included a ledge on which several people were sitting, and above it a large glass window facing up into the pool. The lighting was too dim to make out any faces. Across the room I could see the principal source of light, a wall in which were set large, backlighted transparencies of *Playboy* models in all their natural glory and assorted poses of interest. I moved like a moth toward a flame in the direction of this unique decoration. Oh, well, what was I here for anyway? I might as well admire the principal attraction. It was like having to see the Michelangelo ceiling when you went to the Sistine Chapel, only on a different cultural level.

"What's a boy scout like you doing in a place like this?"

I turned and could just make out her blonde hair.

She moved closer, her eyes on the transparencies.

"They're all good cooks and homemakers, and they want large families with lots of little children of both sexes."

"I don't believe a word of it," I said.

"It's true. That one"—she pointed to a pulchritudinous nymph who more than filled her illuminated rectangle—"is a whiz with apple pie."

"Where do you get your information?"

"I read a lot. And I repeat my question. What's a boy scout doing in this den of iniquity?"

"It's not a den. It's a grotto."

"This grotto of iniquity."

"I'm not a boy scout."

"Yes you are." She put her arm through mine. "You were a perfect boy scout the night you walked me home."

I sipped my drink, trying to find something to say. The pictures were making me nervous.

"Can I get you a drink?" I finally mumbled.

"I have one." She held up her other hand, showing me her glass. "But you can get me something to eat. Besides, it's time to go upstairs. These pictures are making you nervous."

"Does it show?"

"I can't really see in this light. But I can tell. Because you're a boy scout."

"Why do you keep saying that? I wish you'd stop."

She held her face close and looked me full in the eyes. I could see hers sparkling mischievously in the light from the pictures. "I won't stop. I say it because it's true. And I say it because it gets you angry. And you're beautiful when you're angry."

I couldn't help laughing. "Are you ever serious?"

"Only when I do business. But tonight . . . tonight I'm off duty. And I really am hungry."

We worked our way up the narrow winding staircase past couples racing down the treacherous steps. I put my arm around her waist to guide her. When we came upstairs I started to remove it.

"You don't have to let go," she said. "If I don't want it, I'll say something like 'Unhand me, sir.' Then you'd better let go."

"Suits me," I replied. It certainly did suit me—walking into a room filled with book people, with my arm around Dinah Foxworth. I tried to think of it as walking into a room with an attractive blonde, but I couldn't forget that she was a celebrity in our business.

"I guess I'm embarrassed. You're rather a famous person."

She halted and faced me. "It's time you stopped being a boy scout." And with that she kissed me. Then, with her cheek to mine, she whispered in my ear, "Care to look around and count how many people are watching?"

"No," I replied, "I'd rather hold this pose. Just think, tomorrow I'll be an item for all the gossip-mongers."

She broke away, smiling. "You already are an item, sweetie."

"Because of this?"

"No, not this." She pulled me into a small side room loaded with plates and silverware and platters of food.

"What, then?"

"Your walk this afternoon." She began poking into the platters and piling food on her plate. "Farkas told me all about it."

"Mervin told you about the walk?"

"Me and half of Chicago. He's been rhapsodizing about it to anyone he can buttonhole. Here, fill up your plate."

"I'm not hungry."

"Then let me." She began piling food on the plate she had shoved into my hand. "There's more here than I can fit on my plate."

"What did Mervin say?"

"How do I know? I'm not interested in architecture. Hold your plate steady. Those funny little meat things are rolling into the macaroni."

"But you said he was rhapsodizing."

"Look, he made it sound great. And he thinks you're terrific. Is that what you wanted to hear?"

"Did he really say that?"

She propelled me over to an empty space in the large main room and sat down on the arm of a sofa. She seemed preoccupied with eating each tidbit in strict alternation, from my plate and then hers.

"It's very important how you mix these things."

"Are you purposely ignoring my question?"

She looked up from the plates for a moment, then resumed her alternate picking. "If you're thinking of leaving Redwood, that's laudable. I can't understand how anyone could stay employed by that maniac. But if you think you can parlay one afternoon's walk into a job with Farkas, forget it. On the other hand, he's probably been very helpful to you already on that score. Half the people in this room know for a fact that you're a bosom buddy of three of the top editors-in-chief in the business."

"That's idiotic."

"So are half the people in this room, when it comes to believing something. If the right person says it, then it's so." She stood up and handed me her plate. "You don't believe me, but here, get rid of these and I'll prove it to you. First, give me a napkin."

I handed her a paper napkin and she wiped my lips. "Lipstick," she remarked. "I don't want any distractions in this demonstration."

"I was saving that kiss as a souvenir."

"I'll give you another one later. Just try to look intelligent." She surveyed the room while I dumped our plates; then she pointed to a tall woman talking to a small group of people. "Do you know her?"

"No."

"Mona Packenham."

"I know the name."

"Fine. She'll do."

Dinah put her arm through mine and steered us over to the conversationalists. I could see Mona Packenham monitoring our approach.

"Hello, Mona." Dinah's smile was positively radiant.

"Dinah. How nice. What a crush."

"It's hideous. Oh, do you know Howard Miller?"

"No, I don't think . . ." Then the name registered. "Oh, of *course*. Howard Miller. Delighted. I don't know how we haven't run across each other before." She introduced us around, mumbling the other people's names, but repeating Dinah's and mine each time. By the fourth introduction she had adopted a proprietary tone, as if she were showing off old friends.

"What a wonderful idea," she said, "that walk. And here I've been thinking there's nothing to do in Chicago during the day."

"Are you going to lead another walk tomorrow?" one of the others asked. Already I had been promoted to leader.

"I hear we're going to New Orleans next year," said another. "Surely you'll be conducting a tour there."

"You've got to do it again the next time we're in Chicago," Mona declared. "I'm sure you'll have fifty people. Just remember, I'm signing up for it right now."

The others all murmured agreement. We chatted for a few minutes and then Dinah pulled me away.

"Want to try again?" she asked as we pushed our way through the crowded room.

"No. I believe you. And all because Mervin's been talking it up enthusiastically." I shook my head.

"Because Farkas is who he is. Reputation means a lot in this business. If enough agents have heard about that walk, you'll be inundated with manuscripts next week."

"That's an awfully cynical remark."

"I think this party is beginning to depress me."

"Are you sure it isn't my sudden notoriety? Taking all the limelight away from you."

"Oh, it's marvelous how quickly success goes to the heads of some people."

"Would you like me to take you to another party?"

"I thought this was the only party in town tonight."

"It's the only party that people like you are invited to. But the little people—like me, before I became a celebrity—have to have their fun too. Would you like to have fun . . . instead of this?"

"Not with you, if you're going to be insufferable. What kind of party is it supposed to be?"

"It's in an amusement arcade. All the games and pinball machines and things are rigged so they're free."

She considered it for a moment and said, "Why not?"

But it took us almost fifteen minutes to work our way out of the room. We were constantly stopped by someone or other who had to say hello to Dinah or to me and make some remark about architectural tours. Outside in the street, a taxi was pulling up. We grabbed it as the couple inside was getting out. The man was someone I knew vaguely.

"Hello, Dinah," he greeted. "How're you, Miller? How's the party?"

"Noisy," we chorused.

As we climbed into the cab, I heard his companion ask, "Who's that?"

"Dinah Foxworth," he replied, "the sub-rights lady. And Howie Miller. He's a big buddy of Mervin Farkas."

"Ooh," she cooed. "This is going to be neat. You know everybody."

I found Tony Hunt's card and gave the driver the address. Dinah was giggling. "Ooh," she mimicked. "This is going to be neat. I knew I should stick with you."

As we drove through the nearly empty streets, I settled back and put my arm around Dinah. She nestled closer and I decided to risk a kiss.

"What's that?" she asked.

"It's called a kiss. You said you owe me one. Remember?"

"If I owe you," she remarked, "then I have to give you. If you take, mister, that's not the same thing."

"I'm sorry."

"Don't be," she replied. "I'm in a giving mood." She reached up, cupped my face in her hands and kissed me warmly. "Now we're even. Except I give better kisses than you do."

"Well," I replied, "that was my boy-scout kiss. How about if I tried a different style?"

"Some other time. I happen to like you the way you are."

I settled for that.

The arcade Tony Hunt had taken over was gaudy, frenzied, and mobbed with people rushing from one machine to another. The air was filled with noises: pinging, clacking, gonging, and buzzing. Lights were whirling all around us as we entered. It seemed almost as crowded and noisy as the place we had just left, but with a happier, more open feeling. Tony waved a big hello, came over to shake hands with Dinah, and then ushered us over to an unoccupied pinball machine.

"Have fun," he shouted over the din.

And that we did for the next two hours. I stuck mainly to solo games or tests of skill, noting almost at once that Dinah was a fierce competitor at anything that pitted us against each other.

"What's the matter, boy scout?" she crowed. "Are you a sore loser?"

"I don't like overly competitive women."

"That figures. Did you learn that from Redwood?"

"No. From my den mother."

"You're a male chauvinist pig. I should have known."

Of course, she was just as competitive at solo tests of skill, trying to get the best score possible. I came up to her while she was busy with one of these and put my arm around her.

"Unhand me, sir."

"Why?"

"Because if you spoil my aim, I'll whack you in the crown jewels."

"That's a good reason," I said, releasing her quickly.

Finally, at some point past midnight, I could see that she was exhausted. She wouldn't admit it, of course, but simply complained, "These games are rigged."

"Of course they're rigged. Have you paid so much as a nickel to operate any of them?"

"I don't mean that kind of rigged. I mean you can't win at them." She glared at me, but lost it in an unladylike yawn.

"Maybe it's time we headed back to the hotel."

She didn't argue, just waved a friendly goodbye to Tony and walked to the door. When we were out on the street, she put her arm in mine and said, "It's close enough to walk. At least I think it is. Would you like to go into your act again and rescue a fair maiden?"

"If you mean walk you back to the hotel, I hardly consider it an act."

"You're right. It's getting to be a habit."

"I was thinking more in the line of pleasure."

We walked contentedly and silently in what I hoped was the right direction to the hotel. As it finally came in sight, she yawned.

"I think I'm going to hit the pillow and go right out."

"Lucky you," I responded wistfully.

"Why? Are you an insomniac? It doesn't sound like you."

I explained about my arrangement with Milt.

"That's ridiculous. What are you going to do?"

"Wander the streets . . . sit in the lobby . . . I don't know."

"You'll do no such thing. It's criminal the way you let people take advantage of you."

"What do you suggest?"

"I have a double room. There are two nice big beds." She looked at me severely. "And they're four feet apart."

"I can't accept an offer like that. Even if the beds are four feet apart. What if someone sees us?"

"It'll give them something to talk about."

"With my luck, there'll be a television crew in the corridor just as we get to the door of your room."

"Wonderful. I've never been on television."

"Be serious, Dinah."

"I am serious. I wouldn't be able to sleep, thinking of you wandering the streets. Besides, you phony, when was the last time a lady invited you to spend the night in her boudoir?"

"Do you really have a boudoir?"

"A boudoir is simply a lady's private room. It comes from the French for a place for pouting."

"Are you going to pout?"

"Only if you don't come up with me."

That seemed to settle it, and by this time we were in the hotel lobby. I was so obviously uncomfortable and nervous as we got into the elevator that Dinah was overcome with a fit of the giggles.

"Stop that!"

"I can't," she managed to mutter through her giggling. When

the elevator doors opened at her floor, she stuck her head out in an exaggeratedly conspiratorial manner. "Coast is clear," she hissed.

I followed her down the corridor.

"Where's the television crew?" she asked.

"Never mind them. Where's your key?"

Between giggles, she fished into her purse and produced the key, which she held out for me.

"Maybe . . ." she said as I opened the door, "if I get undressed out here in the hall . . ."

I was inside the room, and reached out and yanked her in behind me. She let out a yelp. I closed the door and held her close. Slowly she calmed down.

"Oh, Howard. I really am sorry. But you were so funny. You looked like all the little boys who ever got caught with their hand in the cookie jar, all rolled into one."

"I'm glad I amuse you."

"You really do, sweetie. You really do." Her expression became serious. "Now look! One kiss—one terribly, terribly sexy kiss—and then you lie down over there and stay put." She pointed to the bed farther from the door.

It was a good kiss; I don't know about the sexy. Then I walked over to the bed, removed my jacket, tie, and shoes, and flopped down on top of the bedspread.

Dinah wandered around for a bit, removing her earrings and necklace, selecting a nightgown, pulling down the covers of her bed, all the while humming softly to herself. She turned off the top light and put on a lamp near her bed. My eyes were refusing to stay open. I watched her go into the bathroom from under half-closed lids. A few minutes later she came out in her nightgown. My eyes opened wide.

"Don't get any funny ideas during the night," she cautioned, climbing into bed.

"Maybe you'd like to sleep with a sword between us," I mumbled, "like Siegfried and Brünnhilde."

"How perfectly Redwood of you," she said, turning off the light. "Anyway, you didn't bring your sword. And I lost mine."

"I have a nail file," I volunteered.

"You know what you can do with it."

"Good night." Her voice came across the darkness in a lazy yawn. "Pleasant dreams."

"You bet," I murmured and fell instantly asleep.

I awoke with the first light of morning, slipped on my shoes, and went into the bathroom to wash my face. Dinah was sleeping soundly. I came back into the room quietly, slung my tie around my neck, and put on my jacket. Then I tiptoed over to Dinah's bed and bent down to kiss her. She opened one eye warily.

"Watch out for my sword."

"All the time," I laughed, and kissed her. "Thanks again."

As I slipped out the door I heard her mumble, "See you on television."

Down in the nearly deserted lobby, I used the house phone to call Milt.

"Can I come up now?"

"What time is it?"

"Did I wake you?"

"Of course you woke me. What time is it?"

"Six-thirty."

"You bastard."

"Can I come up?"

"Yes, you can come up. I hope you break your neck on the way."

But he didn't ask where I had spent the night. Even after we had showered, dressed, and breakfasted together, he didn't ask. In return, I didn't ask how—or if—he had made out. For a nosy, gossipy bunch, publishing people can be remarkably uninquisitive on some occasions.

5

By the middle of the week, I could put the ABA behind me, but not thoughts of Dinah. Her schedule, however, was hectic enough that she could put me aside. Out of sight, out of mind, I fretted. At least there was the telephone, but that too was something of a bust. The telephone, she informed me, was an instrument of business for her, not one of pleasure. It was her umbilical cord to the book world outside her office. I had to content myself with longing for her, which, as it happens, suited the Graham Greene mood into which I was rapidly plunging myself.

Visions of Dinah became visions of Dinah framed against mosques and minarets, then just visions of minarets, and finally the streets of Istanbul with shadowy figures pursuing each other relentlessly. Her normally crowded life and my upcoming trip combined to put Romeo and Juliet on hold for a while. I spoke to her from time to time, and we agreed to get together after I got back from Istanbul. Good. I had something to live for, something to keep me going, to survive my secret mission. And if the KGB caught up with me in some dark alley behind a mosque, I would expire with her name on my lips.

I wasn't really a basket case. This fantasizing only crept up on me in odd moments—while shaving, or on the subway, or sweating over a dull manuscript. As a matter of fact, I slept well every night. I even went to another Mets game. And in the office,

Emory assured me at every possible moment that everything would be all right and that he was taking care of all the details.

Emory was true to his word in every respect. For one thing, I was given my instructions over and over again, at least a dozen times. I repeated them back to him until he was as satisfied as he could ever be. For another, he, or Fran Bishop, took care of all my travel arrangements. I was booked on a Pan Am nonstop flight to Istanbul for a Thursday in mid-June, and a return flight on Tuesday. I had a room at the Hilton—on the Bosporus side, whatever that meant—for Friday through Monday nights. And Fran even supplied me with a number of travel folders. I knew that Emory's largesse had its limits, so I had bought my own guidebook to Istanbul earlier. I would be buying souvenir guidebooks for an entirely different purpose over there, and they would, presumably, be replaced by the Kuzatov manuscript, so I needed one of my own. After all, I would be doing a lot of sightseeing in and around Istanbul. I kept thinking of it as my "cover."

I had two copies of the restaurant address, which I memorized in my best spy manner. The restaurant became the "drop" in my rapidly overheating imagination. Every time I thought about it, though, I lost my appetite.

It was decided that I should announce my rather sudden vacation plans casually, which was a neat trick. Only last year I had publicly agonized for weeks about whether I should go to the Cape or Montauk. But Fran covered for me by informing everyone that she needed to know at once what people's vacation schedules would be, so that she could organize the office work load. This was something of an exaggeration, but it served to throw everyone into enough of a panic to dull curiosity about my own uncharacteristic plans.

Most of all, interest in my trip was deflected by Emory's announcement of his own visit to Paris. He made the announcement at a Wednesday editorial meeting. We had been going around the table in rapid order; nobody had much to discuss, and Emory was noticeably impatient.

"If there is nothing else"—his hand came up in a staying gesture—"I have something to say to you all." We settled back in our chairs apprehensively.

"This is a business of rumors in a city of gossips," he began. A smile crossed his face briefly. "Therefore, I am happy to be able to satisfy your hyperactive curiosities and give exercise to your wagging tongues. I have a bit of very important news. It may even bear some relation to rumors that seem to be enlivening the luncheon circuit.

"In about a week's time I will be flying to Paris to appear at what one hopes will be a highly publicized news conference. There, with my friend Itzhak Pierre, of the esteemed I. Pierre publishing house, I will announce the American edition of a major new book by Vasili Kuzatov. Redwood Press will be publishing here virtually simultaneously with I. Pierre. The announcement will be made in Paris because that is where the manuscript is being received."

"Being received?" The interruption came from Myra.

"I need hardly remind you, Mrs. Palmer, that Vasili Kuzatov is looked upon less than kindly by the leaders in the Kremlin. He is widely referred to, even in the popular press of this city, as a prominent dissident. Any book that he might care to write at this time would not be greeted by Moscow book publishers as a big hit with the Politburo. Therefore, it seems wise to publish elsewhere. Fortunately, he is fluent in French, and so a manuscript—"

"Is being received in Paris," Myra interrupted again.

Emory's eyes narrowed. "It is not being mailed as a Candygram, Mrs. Palmer, if that is the drift of your question. You will excuse me, I am sure, if I reserve to myself the exact details of how the manuscript is to arrive in Paris."

The unspoken letters "CIA" hung like a pall in the room's humid atmosphere.

"Needless to say, this is a book of the first magnitude. A major publishing event." Emory paused. "Of course, I have not read it, but I have been informed by reliable sources of its contents." He stressed the word "reliable" in a way that punctuated the unuttered "CIA" in all our minds. "We do know that it describes, in the most distressing particulars, how the Soviet hierarchy systematically denies the process of creative expression in Russia today."

His words were pretty much the same as those he had used to

describe the book to me. This was the fourth time I had heard the speech, and I was struck by how pat and rehearsed it sounded. It occurred to me that Emory really didn't know too much about the actual contents of the book.

When he finished his little set piece, there were questions from everyone. I had been asking most of them myself, in my sessions with Emory. These, by the way, had been held at off-hours, so as not to arouse curiosity. The answers brought none of us any nearer the truth about what was really on our minds: who was smuggling out the manuscript, and how; why Redwood and I. Pierre were publishing it; how it was being authenticated; how Emory, even vaguely, knew its contents. These remained points to ponder, or better yet to skip over, because no meaningful answers were forthcoming.

The meeting broke up when Emory decided to answer one final question.

"When are you going?" Milt Foster asked.

"A week from Friday," Emory replied. "The announcement will be made at I. Pierre's office a week from this Saturday."

On the way out of the meeting, Milt put a hand on my shoulder. "Gee, fella, you'll miss all the fun. Won't you be on vacation then?"

"Yeah." I winked at him. "But our fun's over already. Somehow it was more fun guessing than knowing."

He nodded. "You're right. It's Emory who has all the fun, now."

There was last-minute shopping to do, and packing, and of course another briefing from Emory, but I set aside time for something that had become very important to me. I wanted to discuss the whole trip with Hartley Dobbs. Emory had been very explicit about secrecy—complete, utter, unequivocal, zipper-your-mouth secrecy. Although I could understand his position, in one lunch and a few subsequent phone conversations, I had developed an unexpected attachment to Dobbs. Perhaps it was the attentive way he listened to me. And there was something else, something I couldn't articulate. Dobbs put it in words for me.

"You lead a solitary life," he said. "Not necessarily a lonely one, but one that is alone. And now you are embarking on an adventure, indeed a secret and possibly hazardous journey. It makes you feel all the more alone. You want a confidant. I am flattered that you have chosen me."

We were back at the restaurant for an early dinner. I had chosen the hour, and Dobbs had fortunately agreed to it, because we would be the only customers and have an hour or more for what had to be a private conversation.

I was there when he arrived, nursing my brandy and soda. Franco materialized as Dobbs sat down.

"Martini, extra dry, straight up," Franco murmured.

Dobbs nodded appreciatively and smiled.

"The trouble with this place," he remarked when Franco was out of earshot, "is that you couldn't change your drink if you wanted to."

I thanked him for coming.

"It was your tone of voice, my boy. The prospect of a delicious meal, the suggestion of some important, secret information had nothing to do with it. Your tone convinced me that I was needed. It is so long since I have felt *needed* that I could not resist the invitation."

"Well, I do appreciate your coming."

"Enough. A profusion of thanks is unwarranted. It suggests I'm doing you a bigger favor than I am. After all, I enjoy being noble on the rare occasion, and I will probably enjoy even more the confidences you are aching to divulge. So you see, it's my wants we're catering to, not yours."

I laughed. "You really are the kind of person one is willing to spill secrets to."

"It's been known to happen," he said noncommittally.

"Emory has sworn me to secrecy," I confided. "Knowing him, I'm surprised he's even let me out of his sight. But I want to tell you about what's been happening. So to hell with Emory's vows of secrecy."

Over drinks and dinner I poured out all the details in what I hoped was a coherent sequence. He interrupted a few times to ask some pertinent question or to take me back over some detail, but I did most of the talking. I finished while we were sipping after-dinner brandies, and that was when he remarked about my wanting a confidant.

"That's true," I said. "I guess I've been bottling up all sorts of feelings. But I think I need something more. After all, I could have confided in Paul Ostrow. He knows all about Istanbul; he was going to go there for the manuscript before I butted in and got to replace him. What I really want is someone with answers, and I don't think Paul would have answered me any more than Emory does."

"That may be," Dobbs replied, "but what makes you think that I'll have answers for you?"

"At least you're an outside observer. I'm living with this crazy situation. It confuses me. Do you see something that I'm missing?"

"Not really. Crazy or not, I'm willing to accept the situation at face value. And you seem to have done the same on an instinctive level."

"How's that?"

"You haven't refused to go. And I don't think that's because you're confused. It's because you've accepted Redwood's story."

"Not all of it."

"All of it that counts. Look at it this way: you may be apprehensive about the business in the restaurant in Istanbul, but you don't really doubt what will happen. What I mean is, you don't really expect to be kidnapped by KGB agents. Or do you?"

"The thought has crossed my mind."

"Even so, it simply confirms that you believe what Redwood has said about a manuscript being delivered to the restaurant in Istanbul."

"Do you think it's dangerous?"

"Stop working yourself into a lather. That's a silly question, and you know it. Crossing the street in New York is dangerous. Perhaps crossing the street in Istanbul is more dangerous. I don't know; I've never been there. Eating in a restaurant in Istanbul may not be any more dangerous than partaking of that greasy cuisine anywhere. What can be dangerous is making yourself into a character from Graham Greene or Eric Ambler. Are you planning to wear a trench coat throughout your stay in Turkey?"

I smiled apologetically. "Do I sound that bad?"

"You sound apprehensive at times, for which I don't blame you, and downright silly on occasion."

"Like now."

"Whenever you let your imagination get the better of you. If I were you, I would stop dwelling on the cloak-and-dagger aspects of this unexpected vacation. Instead, I would concentrate on the one truly remarkable point in your entire story."

"What's that?"

"Your employer has offered you, at his expense, a costly and exotic vacation—and this from a man who, to hear you tell of him, ordinarily doesn't give away ice in the winter."

Buoyed by my conversation with Dobbs, I managed the next few days in more cheerful style. I even began to look like someone about to go on vacation. The day before my departure was editorial conference day, but I was hoping to avoid that customary hassle. No such luck.

"I don't suppose you have any business to bring up," Paul said from my doorway, "but Emory wants everybody at the conference today."

Dutifully, I joined the others, most of whom were carrying manuscripts or papers of some kind. Everyone obviously expected Emory to be in a good mood as the day of his big announcement neared. Whenever Emory was expected to be in a good mood, the amount of material introduced at an editorial conference increased dramatically.

As it turned out, he was in one of his businesslike moods, which was often one step from being short-tempered. He wanted everyone on hand so he could go over his press statement for Saturday.

"I want your comments," he stated, "but helpful comments, not frivolous remarks. If you do not understand something, if it is not clear, say so. That would be a helpful comment. If you simply disagree with something I say, that would be frivolous."

Myra made a sound like a suppressed snort. Emory glared at her, and everyone else shot nervous or warning glances in her direction. We were all caught off guard by having Emory monopolize the opening of the meeting, and this tended to make us nervous.

The reading of the press release, punctuated by scattered comments that were presumably helpful, took more than half an hour. It did nothing to lighten the atmosphere in the room. When Emory finally decided that he was finished, he glanced around the table. In front of everyone but me was a pile of books, manuscripts, papers, jackets, printouts, and the like.

"Does anyone else have anything he or she wishes to discuss?" he asked curtly. "If not, I am extremely pressed for time, and the editorial conference is over."

Paul started to push back his chair, but the rest of us sat motionless. I gazed across at Sherwood, Paul, Myra, and Mr. Snap, trying to gauge their reactions. Paul was obviously taking Emory's hint. The others were either stunned or weighing the chances of bringing up a project under unfavorable circumstances. To my right, Minnie began fidgeting in her chair like a bird ruffling its feathers.

"Emory," she squeaked.

He turned to her, his face expressionless.

"Emory," she squeaked again, then gulped audibly.

"That's me."

"The Audrey Burbage manuscript . . . it's been . . ." She blinked her eyes uncontrollably, her face grew red. "You've had it on your desk . . . on your desk for weeks."

"I am aware of that."

"Please, Emory . . ." she gulped. "The agent is pressing me."

"Tell the agent it is not you who is holding the manuscript."

"I did." Now she seemed near tears. "I did . . . I can't . . . I mean . . . I must have an answer."

"Must?" He was pointedly ignoring her extreme discomfort. It was one of his more cruel performances.

"Today," she gasped.

"You must have an answer today?"

Minnie nodded vigorously.

"Then it is 'No!'" He looked at her without expression for another moment, transferred the look to each of us in turn, nodded his head, and marched out of the room.

The conference was definitely over.

Much of the rest of my day was filled with the last-minute fussing that precedes a week's absence from the office, so I paid little attention to *l'affaire Minnie* that was brewing. What I saw of it came in passing bits and pieces.

On my way out of the conference, I had heard Myra snarl, "That bastard!" and something else I didn't catch. And I wasn't surprised to see her passing my doorway, back and forth, all the rest of the morning.

"Poor Minnie," Milt said just before lunch. "By the time she got back to her desk, Emory had already dumped the manuscript on it." He lounged in my doorway, but I fussed away at my desk. "Oh, well," he added, "it'll be a good week to be away."

After Milt walked down the hall, Sherwood appeared in my doorway. "Did he tell you?"

"The manuscript was already on her desk."

"No, no, not that." He glanced right and left hurriedly. "Myra is roaring around the office, trying to whip up anti-Emory sentiment."

"Well, she's avoided me so far."

"No good will come of it," he warned. When I ignored him, he sighed, "No good comes of anything," and wandered off.

I went to lunch alone, catching a glimpse of Minnie and Myra on the street. Myra was hunched over Minnie's little pigeon figure, patting her shoulder and talking to her vigorously.

When I returned, I looked in on Mr. Snap, who was wrapping his damp coffee grounds in some plastic.

"What made you pick Istanbul?" he asked, concentrating more on his limp package than on me. It was the first time anyone had asked me, but I had been expecting the question anxiously for over a week.

"I wanted something totally different. Istanbul seems as different from my usual vacation as I can possibly get."

He nodded and looked over toward Myra's office. "Can you take the Wicked Witch of the West with you?"

"What's going on?"

"If she doesn't shut her big mouth, Emory is going to boot *her* all the way to Istanbul."

"She seems to have avoided me so far."

"Lucky you. She's tried to enlist everyone else."

"To do what?"

"To get Emory to change his mind about that silly woman's book."

"You mean Minnie or Audrey Burbage?"

He gave me a nasty look over his granny glasses. "I told her I don't interfere in editorial matters."

Paul Ostrow stopped by about an hour later. He hung in the doorway and I studiously concentrated on some papers on my desk.

"Emory will not change his mind," he said finally.

"She hasn't been in to see me," I replied.

We stared at each other for a moment, and then he walked away.

It was after three when Myra came into my office.

"I was beginning to feel left out," I said.

"You seemed to be busy with getting away. Are you interested in helping me?"

"Helping you or Minnie?"

"Minnie doesn't seem to want to even help herself." Myra's face showed her frustration. "Nobody seems to realize that Emory's arbitrary display of bad manners is also an exercise in bad editorial judgment."

"It's his company."

"That remark," she snapped, "seems to be on everyone's lips. Thank you for at least being consistent with your fellow sheep. I hadn't really expected any better. He keeps all his little editors in line by throwing them an occasional bone. You have your Hartley Dobbs book, and so Emory can do whatever he wants with Minnie's projects."

In a sense she was right. But the practice wasn't exclusively Emory's. I had heard enough stories from other editors in other houses to know that it was the same everywhere. Editors were kept happy by being allowed their pet projects; and being kept happy, unhappily, meant rarely sticking their necks out for their fellows.

Emory came in late in the afternoon, but Minnie wasn't on his mind.

"Is everything ready on your end?"

"I'm ready, if that's what you mean."

"That's what I mean." He seemed preoccupied. I waited for him to say something else, perhaps to run me through my paces one more time. But he simply stared past me out the window.

"Is everything ready on your end?" I asked facetiously.

He continued staring out the window.

"Itzhak hasn't received the manuscript yet," he finally mumbled.

A chill ran through me.

"Do you think anything has gone wrong?" I asked hesitantly.

"No, no." But he was uncharacteristically subdued.

"You're worried."

Suddenly he smiled. "What—me worry?" His grin broadened. "It's all a bit like *Mad* magazine, isn't it?"

I nodded and returned his grin.

"Everything will be all right. Just follow your instructions. I'm sure you will do nobly. And I'll take care of Paris."

He shook my hand, which made me feel as if I were going off to face a firing squad.

"Have a good trip. And be a good boy. Everything will work out just fine. Istanbul will be one of your fondest memories, and I will be at Kennedy Airport to greet you on your return."

It was a nice closing sentence and was certainly meant to reassure me. Unfortunately it didn't. And as it happened, he was wrong on both counts.

6

Every brochure I had on Istanbul contained a photograph of the sun sinking brilliantly behind the silhouetted minarets of a mosque. It is a soul-stirring picture, the kind that lures travelers to distant cities, and I decided that I had to take one of my own just like it. Well, that picture is either taken in December or from a submarine in the Sea of Marmara. In June the sun sets miles beyond the city's skyline. Not getting that sunset picture epitomizes for me the disappointment of Istanbul. Very little in that grubby city is as picturesque or romantic as the expectation of it.

Coming into Istanbul from the airport has to be one of the most dismal travel experiences on Earth. The trip in from any airport, I've found, is pretty bleak, involving either endless views of salt flats and fields of rubble, or Dickensian slums. But Istanbul has other cities beat. For a good deal of the way, you ride past tanneries, which have been located outside the city proper for a particularly good reason—the smell from them is abominable. Along the way, your eye is caught by a peculiar architectural feature in the hillsides: thousands of shanties, most of them just scraps of wood or cardboard, occasionally odd pieces of corrugated metal. These scarcely habitable but thoroughly populated hovels are dug into, scooped out of, or simply stuck onto a miles-long structure of crumbling stones. Only after a few minutes did I realize that I was looking at the once-fabled walls of Byzantium.

Byzantium, Constantinople, now Istanbul. Its history, its reputation, its very names were colorful almost beyond imagination. Nothing could be more antithetical than these dismal sights. What I didn't realize at the time was that the drive in from the airport is a fitting introduction to the city.

I was still euphorically expectant by the time I arrived at the hotel, which certainly lulled any suspicions I might have entertained. The room was luxurious, and my first view out the window was romantic far beyond my expectations. That's Asia, I kept telling myself, and this knowledge endowed the otherwise ordinary vista with an exotic quality. Now I understood why a room on the Bosporus side was a prized accommodation.

Emory's schedule called for an early evening, to get over the effects of a very long flight, and I tended to agree with his plans. I was excited and probably even more apprehensive than I had been back in New York. But the plain truth was that I was bushed. I ate dinner in the hotel restaurant, which was every bit as luxurious as the brochure promised, and went to bed early. In a way, I found it reassuring to conform to Emory's schedule, and even more reassuring that the schedule seemed so sensible.

What I did about sightseeing was to sign on with the most comprehensive guided tour of the city on the first day, and plan to go back to those places I liked the following day. A day's tour of Istanbul is a full day indeed, even without the usual lunch stop at the tour guide's brother's restaurant. The idea of catching a glimpse of everything on the first round and going back to favored spots, a routine I've learned to follow in most cities, has an added advantage in Istanbul. The first day I loaded my camera with slow-speed daylight film for exterior shots, and the second day I went with high-speed daylight film for shooting interiors in available light. As far as I'm concerned, certain interiors are what make Istanbul worth visiting—if anything does.

The Blue Mosque, for instance, has one of the most beautiful interiors I have ever seen. From the outside, one would never guess at its breathtaking splendor. The Topkapi Palace, for all its magnificent location, becomes a worthwhile attraction only after you begin wandering through its countless rooms. The Sulemaniye Mosque, the most impressive from the outside, is even better once you're inside. And St. Sophia, definitely the oldest-looking and dirtiest major building I have seen, is somewhat less depressing inside. As for famous exteriors, the Golden Horn, a location that by name alone conjures up visions of the Arabian Nights, is about as exotic as Hoboken harbor.

I had dutifully made an early evening of it on the day of my arrival, and I was just as dutiful about making a night of it after my day of sightseeing. There had been signs at the Topkapi announcing a performance of Mozart's *The Abduction from the Seraglio* that evening. What could be more fascinating than to see the *Abduction* performed in an actual seraglio? It would be like seeing *Aïda* at the Pyramids. Sometimes I think my mind has been permanently warped by the four years I spent as an advertising copywriter. Lured by the blandishments of my former trade, I taxied back to the Topkapi after an early dinner. The performance was presented not in one of the harem rooms, but in a garden, an open space that proved to be an acoustic black hole. The scenery was on a par with that of a high school production. And the performance, sung in German by one of the women and in Turkish by everyone else, was on much the same level. So much for Turkish nightlife.

When I stepped out of the front door of the hotel the next morning, I was as keyed up as I had ever been in my life. I guess I looked like what I was, an American tourist, casually dressed, with my camera hanging from around my neck. Inside, in my trench-coated view of things, I was something between a secret agent and the last hope of the Western world. Another full day awaited me; I wanted to go back to at least four places I had visited the day before. But my interior clock was set for 6:00 P.M., and every moment until then was a moment in limbo.

A taxi pulled up in the circular drive of the hotel. The driver, a young man, jumped out and opened the cab door for me.

"Where to, good sir?"

A large sign posted at the hotel entrance listed the standard taxi fares to all the major tourist attractions. I checked it, and the driver caught my glance.

"No attention to pay to that, good sir," he said hurriedly. "I am happily yours for the whole day. I take you everywhere your desire is to go. Each place I wait, then take you to next place."

It sounded like a good idea, but before getting into the cab, I asked how much he would charge. He named a figure that I quickly translated into about twenty dollars, which sounded reasonable. Then, just as I was about to step into the cab, I remembered that the guidebook said that haggling was a way of life in Turkey. You weren't supposed to accept any price without arguing it down. I halved his offer. With much indignation he announced that I was trying to rob him of an honest day's income, and then smoothly slipped in a figure equal to about fifteen dollars. It was just like the guidebook's example. I got in and settled myself, confident that James Bond couldn't have done better. The driver slammed the door happily and climbed into the driver's seat. He was probably confident that Kemal Ataturk couldn't have done better.

Haggling was one of the things that bothered me about Istanbul. It went along with an even worse habit. If you stopped to look in a shop window, the proprietor would come out, tug at your sleeve, and try to drag you inside. Then the next-door shopkeeper would come out, tug at your other sleeve, and try to drag you into his shop. The only thing I found worse was that when you walked on many streets, beggars crouching on the sidewalk tugged at your pants leg.

The arrangement with the taxi driver was a convenient one. It allowed me to get quickly to and from the Blue Mosque, the Sulemaniye, the Topkapi, and St. Sophia. Along the way, the driver kept up a steady flow of tour-guide chatter. About half of it corresponded to what the previous day's guide had delivered. The other half, which turned out to be more imaginative than informational, sent me thumbing through my guidebook. He was disappointed that I limited myself to four places, particularly that I didn't want to go to the Covered Bazaar. I had been there the previous day and found it appalling, the sleeve-tugging capital of the world.

The day passed smoothly, and picture-taking filled my nine-hour limbo. From time to time I thought about six o'clock, but mostly I was occupied with little things. I had to remember to take off my shoes before entering the mosques—they give you slippers to wear—and then to remember where I had left them, and to tip the man who was watching them for me. I had to make sure no person at prayer was included in one of my pictures. And then too, the interior of the Blue Mosque was a magnificent distraction. So was much of the sprawling Topkapi. If there were five hundred tourists there, four hundred ninety-five of them were huddled in front of the showcase with the jeweled dagger that was featured in the movie *Topkapi.* That made everything else easier to see.

At last it was approaching six o'clock. I asked the driver to take me to Taksim Square. He let me out, very conveniently, in front of a book-and-souvenir shop. I thanked him profusely and tipped him enough to bring the total back up to his original figure, for which he thanked me even more profusely. Then he drove off and I entered the shop.

The purchases were easy, three souvenir guidebooks, with multilingual texts and poorly printed color pictures. If, as I hoped, I was going to lose them, they were no great loss. The shopkeeper slipped them into a plastic bag, which he folded over and sealed with a quarter-inch strip of tape. I was ready for the restaurant.

I had no particular difficulty in finding it, but, ever the cautious double-checker, I stopped outside the doorway to make sure. After forty-eight hours in Istanbul, I should have known that was a mistake. Immediately next door was another restaurant, and within ten seconds the proprietor was out, tugging at my sleeve.

"Come into my restaurant, good sir," he cried. "The prices are more reasonable." He began pulling me toward him.

Instantly the proprietor of the restaurant I wanted materialized in his doorway and tugged at my other sleeve.

"No, no!" he said, with great determination. "The choicest morsels of authentic cooking are in my restaurant."

"My restaurant is cleaner," insisted the first restaurateur, taking a firm grip on my arm.

"There is no restaurant in New York that serves Turkish cooking like mine," said the second, also claiming a firm grip.

"All American tourists prefer my restaurant," his rival shouted.

"My restaurant has a roof garden," said the other in a wheedling tone. "The good sir would certainly prefer to eat in a roof garden."

That did it. I nodded and smiled at the roof gardener, and his rival immediately let go of me as if I were a leper.

As I entered the restaurant, cursing Emory for having chosen one right next door to another, I marveled at the speed with which Turks turn on and off their enterprise. They'd been at me in an instant, tugging and wheedling. Both restaurateurs had known at once to speak to me in English. And, rebuffed, the rejected man was off me as if I had the plague, the sooner to pounce on the next passing tourist.

I must say that both of them, in a way, had been fair. The other restaurant may have been cheaper; it certainly must have been cleaner. I can't imagine any restaurant being filthier than the one I was in. And my host's claim, that no Turkish restaurant in New York was like it was true. Any restaurant of its caliber would have been closed by the Board of Health.

We climbed three dimly lit, twisting flights of stairs to the roof garden. It was empty, as best I could tell in the near darkness, and as dusty as the set of an old horror movie. The proprietor led me to a table across which a cat was walking with stately deliberation. He pulled out a chair with a great flourish, waved me into it with one hand, and shoved the cat with the other. As I sat down, I looked around, but the only light was what little could make its way through the skylight, and that hadn't been cleaned since the days of the Ottoman Empire.

"Can I leave this somewhere?" I asked, waving my plastic bag.

"But certainly, sir." The proprietor nodded and bobbed like a marionette on the ends of some tangled strings. He took the package gingerly, smiling and nodding. It had to be the least inconspicuous transfer in the history of espionage.

He tucked the package under his arm and reached out with his other hand for what I took to be my throat. I jumped back in my chair.

"Your camera, sir?"

"No, no!" I said breathlessly. Then, more composed, I smiled and said, "I'll just keep that with me." After all, the fate of

Western civilization is one thing, but I had paid two hundred dollars for that camera, and it still had film in it.

"Your package will be waiting for you at my station itself downstairs," he said with an intensely meaningful leer. "It will repose at all times in my protection. I will guard it like a hawk."

I nodded appreciatively and smiled back at him.

"Does the good sir wish to see a menu now?"

I nodded again and smiled. He nodded and smiled some more and scuttled over to a sideboard. My head is going to fall off, I thought, and roll around this dustbin, grinning like the Cheshire Cat. Graham Greene's spies never smiled so much, I told myself.

He returned with a menu, which he wiped on his sleeve, and handed it to me. "A waiter will take your order," he murmured. I occupied myself with peering at the menu in the gloomy light, fearful of another round of nods and smiles.

In a few minutes a waiter clumped up the stairs and sidled over to the table. I gave him my order and handed him the menu. He took it, scratched his head with it for a moment, then nodded to himself, as if remembering what came next, and removed a dust-covered plate from in front of me. As he wandered off, I stared at the tablecloth. A round white circle, where the plate had been, stared back at me.

The less said about the meal, the better. At least it occupied me for an hour, the amount of time I considered necessary for my stay at the restaurant. When I came downstairs from the roof garden, I looked around for the proprietor and his station itself, whatever that might be.

There were three or four single customers at tables—it was hard to tell in the even gloomier downstairs light—and a waiter, not mine, holding up a rear wall. But no proprietor. Near the front door was a cubicle, somewhat like a hat check in a decent restaurant. The proprietor's station itself, no doubt. It was empty.

I walked over to it and peered in. Nothing. So much for the hawk of Istanbul and his eternal vigilance.

"Can I help the sir?"

I wheeled around. The waiter who had been propping up the wall stood about a foot behind me.

"I was looking for the proprietor."

He stared at me quizzically.

"The owner . . . the boss."

He nodded comprehendingly.

"Is he here?"

Slowly, with much peering, he looked all around the room. Some other time I might have admired his performance, but now I was getting nervous.

"Look," I said, "I left a package with him. A package." I began measuring out a parcel of air with my hands.

Now it was his turn to admire my performance. He studied the empty space between my frantically moving hands, nodding thoughtfully.

"A bag . . . a plastic bag . . . with some books."

"Bag," he repeated, holding up his hands, measuring out roughly the same amount of space.

Then, without indicating any further comprehension, he shuffled past me, reached over the ledge at the front of the cubicle, and pulled out my plastic bag.

"That's it," I said. "Thank you."

He handed it to me wordlessly. It was about the same size as before and sealed with the minuscule strip of tape. I resisted the temptation to examine it, smiled, thanked him again, and grabbed it. Then I gave him my bill with a fistful of money, and, smiling and nodding, I backed out of the door.

As I turned up Taksim Square, I felt the plastic and tried to feel whether it contained three books or one manuscript. My fingers seemed to have lost all sensation. Finally, two blocks from the restaurant, I stopped by a store window. While gazing sightlessly at a display of some merchandise or other, I casually undid the strip of tape. I reached into the bag and riffled the unmistakable pages of a manuscript.

Now I know what is meant by that overused phrase, "relief flooded over me," but it was a short-lived sensation. Uppermost in my mind was the thought that I still had two more nights in Istanbul, a day of killing time, a long flight home, and that goddam manuscript in my possession.

It was a little after seven in the evening, still quite light out and just about the deadest time of day in Taksim Square. I couldn't find a taxi. At most times, taxi drivers would cut across traffic and practically come up on the sidewalk to solicit my fare. Now I couldn't find a single lousy cab. After the restaurateur's disappearing act, which had aged me ten years on the spot, this was all I needed.

Actually, it took me a few minutes to walk around Taksim Square, and in that time a taxi came along. But the bag was becoming heavier every second and seemed to be burning my hand. I began to breathe heavily and to perspire profusely. In a word: panic. All the fears of my trench-coat imagination were welling up within me, and I expected KGB agents to pounce on me momentarily. I don't know how I managed to get into the taxi when it materialized, because the driver obviously had just been dispatched from Moscow and was going to spirit me away to God knows where. I sat bolt upright in the back, with the window open so that any paralyzing gas could be dissipated, and clutched the plastic bag. I had just decided to remove the camera from around my neck and hold it by the strap, the better to swing it as a weapon, when we pulled up at the Hilton.

The first thing I did after locking myself into my room was to place the bag in my valise. The second thing I did was to remove my shirt, which was soggy with perspiration. The third thing I did was to sit on the edge of the bed and stare at the plastic bag lying benignly amid some junk in my valise. At that point there was no fourth thing I felt capable of doing.

Somehow I managed to shower and dress for the evening and even leave the room for a few hours. When I came back, the plastic bag was exactly as I had left it. I opened it and glanced at the manuscript. It was just what it was supposed to be, for all I knew: a manuscript of some three hundred pages, typed in Cyrillic characters, which were incomprehensible to me.

Before leaving the room the next morning, I decided on a stratagem familiar to me from spy and mystery novels. I pulled a hair from my head (which hurt; I don't know why the novels never mention that) and placed it between the second and third pages of the manuscript. Then I resealed the plastic bag and dropped it into my valise, on top of some soiled shirts and next to some souvenirs. After a moment's consideration, I closed the valise but didn't lock it. My instructions were to leave the manuscript as I would some souvenirs, or my camera when I wasn't using it, neither secreted away nor lying around. Satisfied, I locked the door and left for my day out of Istanbul.

The trip up the Bosporus to the Black Sea has its attractions and disappointments, like everything else in Istanbul. One of the attractions, photographically speaking, comes when the ferry passes under the bridge connecting the European and Asian sides. The more obvious sights, the palaces that line the Bosporus, are a disappointment, because they are almost all empty and in a state of disrepair. The biggest disappointment is the Black Sea. Some miles up the Bosporus, the ferry comes to a halt and turns around. A voice on a loudspeaker intones, "Black Sea," repeats it in several languages, and everyone rushes to the rail. More miles ahead than we have traveled, the hills on either side of the Bosporus part and a distant view of open water appears briefly. There is time to get one picture, if you're quick, and the ferry is chugging back down the Bosporus.

On the way back, we stopped for a while at Rumeli Hisar, a hilltop fortress. The magnificent courtyard has been turned into an open-air theater where, in the evenings, Shakespeare is performed. Having sampled Turkish Mozart, I declined another evening of culture, but many of the tourists returned to the ferry clutching tickets and smiling in anticipation. It struck me that there must be some people who actually enjoy Istanbul. There's no accounting for taste.

In the afternoon I took another ferry across to the Asian side and spent a few hours wandering around in the section called Uskudar. It was the subject of a song made popular in the 1950s by Eartha Kitt. Her rendition was delightful, but then she probably hadn't been stopped in the street by a leprous beggar. After my own encounter with one, I had a sudden desire for a hot shower. Besides, the sun was low in the sky, and I wanted my picture of the minarets against the sunset. So I hopped the ferry back, secured a place at the rail, and discovered that I wasn't going to get the shot, because the sun would be setting far to the right of the city's skyline.

By the time I got back to my hotel room, I was ready to pack and get the hell out of this so-called jewel on the so-called Golden Horn. I was hot, tired, and hungry, and had probably been infected by a leper, but I remembered my priorities. Carefully, I lifted the plastic bag from my valise, undid the tape, and removed the manuscript, keeping it face up at all times. With the tips of my fingers I lifted the first two pages. The hair was resting on page three, in just the position I had placed it. I pushed the manuscript back into the bag, resealed the plastic with the minuscule strip of tape, which was now losing its stickiness, and dropped it back in my valise. What I would have done if that hair had not been there, I have no idea.

Istanbul's final indignity for the unwary traveler is its airport. On my arrival I had been whisked through by courteous, efficient attendants. In the intervening four days they had apparently all been deported for interfering with the national character. The sluggish village idiots who had replaced them seemed incapable of understanding anything, especially that they were supposed to be running an airport. The porters had gone on strike, as near as I could figure out, though I could manage my own valise and a flight bag. The clerks knew nothing except that my plane would be late in departing. The air conditioning was either nonexistent or inoperative, and the large, shedlike waiting room had a plastic roof which increased the greenhouse effect on the area it covered.

I sat for over an hour in this Turkish bath, my flight bag on my lap. The manuscript, needless to say, was in the flight bag, along with my camera and canisters of film and some crossword puzzle magazines. From time to time a loudspeaker blared forth messages in a variety of languages. None of them concerned my flight. On the other hand, none of them said that the Turkish police and the Russian ambassador were looking for a CIA agent disguised as an American tourist. That was something.

At last they announced that passengers could board for the flight to New York. I marched to the gate as I would to a scaffold, cleared an inspection so perfunctory as to lead one to assume that hijacking was unheard of in this part of the world, and found my seat on the plane. I supposed at first that they had let me board unmolested because it would be easier to take me in my seat; but once we were airborne I realized that it was simpler for them just to blow up the plane. Who "they" or "them" were wasn't too clear in my mind. Suffice it to say it was an anxious trip home.

I was standing on the customs line at JFK, waiting to have my baggage examined, when I heard the announcement over the PA system.

"Mr. Howard Miller, please report to the Customs Inspector's office."

I grabbed my valise and hung the flight bag from my shoulder and began looking all around me.

"Mr. Howard Miller, please report to the Customs Inspector's office."

"That's me," I said to the woman standing in line behind me.

She edged back, smiling wanly.

"Mr. Howard Miller . . ."

Clutching my bags, I stepped out of line and headed for the front of the room.

". . . please report to the Customs Inspector's office."

A uniformed customs agent stopped me.

"Just stay in line, fella."

"That's me," I shouted, nodding up at a loudspeaker.

He followed my gaze and studied the loudspeaker suspiciously.

"You Mr. Miller?"

I nodded vigorously.

"This way." He prodded me toward one end of the room, pointing ahead to a door. His manner suggested that, now that he was involved in the biggest smuggling capture of the century, he wasn't going to let me get away through any negligence on his part.

As I was being hustled to the door, discreetly marked "Customs Inspector," I decided that this was typical Emory. Leave it to Emory to lay on an official meeting instead of just picking me up at the gate.

"In here." My captor opened the door and pushed me into a small office crowded with a desk, several files, some chairs, and four men. Three of them were in dark gray suits, one was in shirtsleeves, and none was Emory. They all turned to stare at us.

"What is it, Wilson?" the man in shirtsleeves asked.

"This is your Miller guy."

"Thank you, Wilson." He continued to stare at us for a moment, then added, "It's all right, Wilson, we can manage. Mr. Miller hasn't done anything wrong. He has a package for these gentlemen."

Wilson exited grudgingly. I noticed the expression on his face because I was wildly looking all around me, searching for Emory. But I also caught that remark about a package.

"Where is Emory Redwood?" I asked, turning to the others, who were facing me from around the table.

The tallest of the three in dark suits reached across the desk, waving an object I couldn't quite focus on.

"Redwood couldn't make it," he said. "But I'll take the manuscript."

"No!" I shouted, getting more upset with each passing second. "No. I can only give it to Emory."

"He's been detained, Mr. Miller," Dark Suit One insisted, waving whatever it was at me across the desk.

"Where is he?" I clutched my flight bag. "This is his property."

"Mr. Miller, please . . ."

"I want to know where Emory is."

"He's in jail, for God's sake," Dark Suit One shouted back. "There's been a murder."

At which point I finally focused on the wallet he was waving at me, at the identity card he was trying to make me see.

"Murder?" But my attention was diverted from the question by his outstretched wallet.

A small inscribed pin was attached to the wallet, but I couldn't make out what it said. And there was obviously some kind of official identification in the plastic case just beneath it, a card with printing on it and a photo. But I couldn't concentrate on that either. My attention was riveted on the name typed neatly in the center of the card: Myles Harrington.

7

"Murder?" I asked again. I was leaning forward, peering at Harrington's identity card, and must have looked as if I would topple over.

"Perhaps Mr. Miller should sit down."

This was from Shirtsleeves, and Harrington responded by whisking away his wallet and coming from behind the desk.

"Good suggestion, Struthers. I may have given Miller a bit of a shock." He took me by the shoulders and guided me into a chair. "Perhaps a glass of water . . ." he added vaguely, and one of his dark-suited associates left the room. He returned a moment later with a half-filled paper cup, and by that time we were all arranged in the crowded room.

Harrington was in a chair behind the desk. Struthers, the rightful occupant of the office, contented himself with leaning against a file cabinet. Dark Suit Two parked himself along the wall behind Harrington and just to his left. Dark Suit Three, after handing me the water, put his back against the closed door. I was in a chair in a corner of the office, my valise at my feet and the flight bag in my lap. The four of them faced me in a semicircle and patiently watched me sip from the paper cup.

"Feeling better?"

"Yes, thank you." I stood up, placed the empty cup on the edge of the desk, and reached into my flight bag.

"You're Myles Harrington."

He nodded, his eyes on the flight bag.

"I know who you are. . . . I mean . . . well, if Emory isn't here . . ." I handed over the plastic bag. "This is for you."

Dark Suit Two stepped forward, took the package from me, and, at a nod from Harrington, placed it on the desk.

All four of them relaxed perceptibly.

"An orderly and voluntary transfer, Struthers."

The shirtsleeved customs man smiled at Harrington. "Perfectly orderly," he said, "and entirely voluntary."

"Have you any questions of Mr. Miller?"

Struthers crossed over to me, his hand extended. "Did you fill out a customs declaration on the plane, Mr. Miller?"

"Yes," I replied, reaching into my jacket pocket. I handed it to him and he read it carefully.

"Nothing to declare beyond the usual souvenirs?"

"No."

"Did he list the manuscript?" Harrington asked.

Struthers shook his head.

"What were you going to say," Harrington asked me, "if customs questioned you about it?"

"I was going to say that it's a manuscript I took to read on the trip. I'm an editor; reading manuscripts is my business."

Of course, we both knew that the manuscript was in Russian. Harrington just looked at me—sadly, I thought.

"Do you want to examine his luggage?"

Struthers slipped the declaration into his shirt pocket. "I don't think that will be necessary." He tapped his pocket. "All I really need is this."

"We'll need the office a little longer," Harrington said.

"Take your time."

Dark Suit Three opened the door for Struthers, who exited without another word, and resumed his stance. The clicking of the latch behind his back sounded uncommonly loud.

For the third time, I asked: "Murder?"

Harrington gazed at me in his slightly sad but mostly inexpressive way, and finally took a deep breath, pushing it out with a sigh.

"This is a pretty complicated situation, Miller." He rubbed his chin with his left hand and tapped the plastic bag with his right. "Do you mind if I just look at this first? Then, I promise you, I'll answer your questions. In fact, I'll have a few of my own."

He picked up the plastic bag and, as he did so, the tiny piece of tape came unstuck. This seemed to bother him, and he fiddled with the tape a bit, testing its adhesive power.

"Have you been opening and closing this, Miller?"

"Just once or twice. Twice, I think. But it wasn't very much of a piece of tape when I got it . . . when it was first put on."

"When was that?"

"Just before I went to the restaurant."

"That would be . . ." He gazed upward, as if consulting some astral clock. ". . . about fifty hours ago."

"It seems like fifty years."

"Hang in there, Miller." He opened the package and removed the manuscript. After glancing at the title page, he began slowly lifting the pages one by one. At page three the hair slipped out. He put down the manuscript, picked the hair off the desktop, and examined it.

"I put that there," I mumbled, my face flushing.

"How clever." His tone suggested otherwise, but he pushed the manuscript aside. "Your opinion, then, would be that this is the manuscript you picked up at the restaurant, and that it hasn't been tampered with since you picked it up."

"Yes."

"The tape was nothing much to speak of when it was first affixed. And that was where?"

"In a bookstore in Taksim Square."

"Then it was lifted and restuck in the restaurant. And twice by you. When were those times?"

"Once in the street . . . I mean, well . . . right after the restaurant."

"To check the merchandise, as it were."

"Yes."

"And the second time to place the telltale hair."

"Yes."

"Not a third time to check if the hair had been disturbed?"

"Oh, yes, of course. I forgot about that."

"Four times, then, altogether. And a stingy little bit of tape.

I'm satisfied, if you are, that the manuscript has been undisturbed while in your possession."

"Oh, quite." I leaned forward and placed a hand on the desk. Dark Suits Two and Three stiffened, but Harrington remained impassive. "Please, Mr. Harrington, what's been going on? You said that Emory is in jail. You mentioned murder. Please fill me in. Who's been murdered?"

"An editor in your office named Minnie Heffernan."

"Oh, my God!" I shuddered. "Minnie, poor Minnie." An image of her, pigeonlike, seemed to rustle beside me. I shuddered again and began to shake.

"Hang on to yourself, Miller."

"I'm all right," I said, calming down. "It's just that I've been so strung out these last few days. And it's been a long flight. And I kept imagining . . . something might happen . . . to the plane, I mean. And now this. Poor Minnie. What happened?"

"I'm not exactly sure." Harrington rubbed his chin some more as he considered what to say next. "Let's put it this way: our friend Redwood has one idea of what happened and the city cops have another. I haven't made up my mind, mainly because even the facts that have been established don't make much sense."

"Is Emory really in jail?"

"Well, it's more a matter of being held for questioning. For one thing, he found the body. But mostly it's his insistence on . . . well, as I say, he has his own idea of what happened."

"Is there some question? Is he a suspect?"

Suddenly Harrington burst out laughing. "Oh, he's a suspect, all right; but not of murder. The city boys think he's trying to get some publicity. But at this point they just can't be sure. To tell the truth, I'm not sure myself."

He sat there beaming at some joke that was certainly eluding me, then resumed his impassive stare.

"Look, Miller, let me tell you what seems to have happened. But you can help me. As I get to certain points, I'll need your opinion on whether they make sense or not. Okay?"

"Okay."

"Let's go back to last Thursday. You flew out to Istanbul that evening. By that time we had a problem: the French manuscript should have arrived in Paris, but it hadn't. Redwood had mentioned this to you on Wednesday, but we didn't want to worry you, upset you in carrying out your end of the operation, so we let you fly out as if nothing had gone wrong. By the time you arrived in Istanbul, Friday, Redwood was on his way to Paris, ready to tear the place apart.

"It was a waste of time. The manuscript simply had not appeared. Redwood and some of our people were checking on that end. I was checking through Washington. Never mind all that. We were stymied. Everything depended on the delivery of the Russian manuscript to Istanbul and your picking it up. Redwood wanted to fly to Istanbul—not that he didn't trust you, but . . ."

"Emory doesn't believe anything is done right unless he does it himself, Mr. Harrington. It's all right; I'm not offended. But I'm glad I didn't know about any of this."

"That's what I figured. Besides, we had a plan, and we were going to stick to it. The news conference in Paris was simply not held, and Redwood flew home on a flight that got him into New York on Monday. By that time we had word that the pickup in Istanbul had gone smoothly. All we could do was wait for you to return home.

"Now here's where things get a little hairy. I'd appreciate your comments if anything I say doesn't sound like the way things are done at Redwood Press. I'm reconstructing this in my own words, mainly from Redwood's statement."

"I understand."

"Okay. Redwood gets in on Monday and goes to his office. He's in a less-than-happy mood, even though I phone him to tell him that everything seems to have gone right in Istanbul. Less than happy . . . hell, he's in a real snit. I gather, from what he's said and from some conversations I've had, that he was a holy terror around the office that afternoon."

"That sounds right. I'm glad I wasn't there."

"Well, it might have helped if you had been. All he wanted was to get his hands on that manuscript you were bringing in. I know, because I spoke to him about five times between his arrival and midnight."

A thought occurred to me. "Everyone must have been asking what happened about the news conference in Paris."

"Right. And that didn't help his mood any. But he just fobbed them off, he said, with something about a delay."

"That's Emory. Never apologize, never explain."

"Anyhow, overnight he worked himself up into a really monumental rampage. That's not how he says it, but I'm telling you. I flew up this morning and arrived in his office about noon. He was really on a tear. I've seen our friend Redwood in lots of different moods, but this was something new. Of course, he was on his own turf, so he was giving himself free rein. He was running up and down the corridors when I arrived, in and out of offices. . . ."

Here he broke off, and a bemused look seemed to cross his face.

"Miller, tell me about 'over the transom.' "

" 'Over the transom'? It's an expression we use in publishing to describe unsolicited manuscripts. Why? Was Emory going into one of his fits about the pileup of unsolicited manuscripts?"

"That was part of it. Does he get these fits often?"

"Not really." I shrugged it off, feeling suddenly disloyal. "The manuscripts tend to pile up. We're supposed to read them on a regular basis. We don't . . . they begin to overflow . . . and if Emory's in one of his grousing moods . . ."

"Does he usually hand them out? Or do you usually go through the pile yourselves?"

"Was he handing them out? Then he was really pissed off. I can recall him doing that only once before."

Harrington considered this for a bit. "I think I want to get back to this business," he said. "You'll see why in a minute. But to continue . . . I arrived around noon, as I said, and Redwood was running around dumping piles of manuscripts on people's desks. He asked me to wait in his office, which I did, and he held a kind of rump meeting next door in Ostrow's office. Between the open door and his excited tone, I could pretty much hear what he was saying.

"He had Ostrow in there, of course, and the two women. He went on for a while in what I take is his customary manner when he's chewing people out. He can be pretty sarcastic."

"Oh, yes."

"There was one line about how 'over the transom' doesn't mean 'under the rug.' I kind of liked that one. Anyhow, the upshot was that he sent the three of them home in taxis, each with a pile of unsolicited manuscripts."

"He's never done that before," I volunteered.

"I think he had simply made too much of a fuss—he was, after all, in a highly nervous state—and he didn't know how to come down from his high horse. Besides, I was in the office, and I think, if he could have, he would have sent everyone home.

"Once he got that taken care of, he calmed down. But only a bit. There was still the manuscript you were bringing in from Istanbul, and that kept him edgy. I went over with him the arrangements we were setting up for this evening here at Kennedy, and he could hardly pay attention. In fact, by one o'clock, I was glad to leave him and head out here to set up your reception. So I was nowhere around when the call came in and everything fell apart."

Harrington sat back in his chair. I leaned forward in mine. His irritating habit of breaking off or deflecting his narrative at the crucial points was, I suppose, some special technique he had developed. It was draining me, which must have been noticeable, but didn't seem to bother him in the slightest.

"What I tell you now," he went on, "is Redwood all the way. As I explained, I was out here with these gentlemen." He nodded at Dark Suits Two and Three. "I didn't get back to see Redwood until about suppertime, and he was at the precinct for questioning then. So, if something I say doesn't make sense to you—and some of it doesn't make sense to me—please let me know. Just stop me."

I nodded, wishing he would just start.

"About two-thirty, Redwood was in his office when the Heffernan woman phoned him. She was apparently in a highly agitated state, and Redwood could hardly understand her at first. Finally he managed to make out what she was saying. 'Babbling' is the word he used. She was telling him she had the manuscript, Kuzatov's manuscript, the one in French, the one that was supposed to be in Paris. It seems it had been in the pile of unsolicited manuscripts."

He looked at me questioningly, apparently expecting a comment.

"Okay," he said, "we'll get back to that. You can imagine Redwood's reaction. He asked her some questions, had her read a few things to him from the manuscript, and then he was off like the proverbial shot. He arrived at her apartment, he says, by three o'clock. Unfortunately he didn't check his watch either for the call or for his arrival at her apartment. But two-thirty and three are close enough.

"The door to her apartment was open when he got there; not wide open, but off the latch. The cops have been over this with him several times, and he sticks to it. He went in, calling to her. There's a small hallway, the length of the wall of the kitchen on the other side, and then you come right into the main room. Heffernan used this as her living room and workroom when she brought work home. There's a desk in there, near a large window, and manuscripts were piled on it. She was on the floor next to the desk."

Again he paused, this time, I assumed, for dramatic effect.

"At least Redwood kept his head," he continued. "But then, good training always shows. He knelt down to see if she was alive or not, and made sure not to touch anything. She didn't appear to be breathing; there was a bruise on her temple and a large, ornate candlestick lying next to her head. He immediately phoned the police, gave them his name, the particulars of what appeared to have happened, and got the ball rolling. Then he paid attention to what he had come about, what was uppermost in his mind—the manuscript.

"He went through the pile of manuscripts on the desk, moving them as little as possible, searched the floor around the desk and the body, even—he admits—felt under the body. No Kuzatov manuscript. It was gone . . . if it had ever been there."

"Do you doubt it?" I asked.

He considered that for a moment. "On the face of it, and that means accepting Redwood's account, there's not much room for question. A question or two, maybe, about some particulars . . ." He broke off, put his hands behind his head, and leaned back in his chair.

"Let's go back to that 'over the transom' business," he said.

"It's one of the particulars that bothers me. Though I must say it doesn't seem to bother the city boys. Some of those detectives are pretty shrewd questioners, and they've been over the whole story of the two manuscripts as Redwood gave it to them. They've even been over it with me. And it's obvious they buy it. What they don't like is what Redwood is making of it. They think he's looking for a big hoopla, so he can get lots of free publicity for the book."

"Mr. Harrington," I interjected, "you're making my head spin. Maybe it's my training as an editor, but your narrative style is for the birds. Can you just please tell the story straight through?"

Dark Suit Two snorted or coughed, and Dark Suit Three shifted himself uncomfortably against the door.

Harrington ignored them and played with a speck of dust on the desk. "My narrative style," he murmured, examining his fingertips.

"I don't mean to be critical," I went on. "But I've had a long and harrowing flight home; nothing has gone right since I landed; I find out a colleague has been murdered; my boss is being held for questioning; you're asking me all sorts of questions . . . and all I can think of is . . . Minnie." My voice broke.

He jumped up. "I'm sorry, Miller. I wasn't thinking. Would you like something stronger than water?"

"No. I'm all right." I took a deep breath. "Just please tell me what happened next."

"Are you sure?"

I nodded, and he sat down again.

"Redwood," he said, as if there had been no interruption, "had about ten minutes before the cops came, to think over his story. They figure he came up with his bright idea in those ten minutes, and they're probably right. Anyhow, after answering all their initial questions, he began insisting that the KGB had done it. That they had found out about the manuscript . . . had broken into the apartment . . . whacked the Heffernan woman on the head . . . and stolen the manuscript. This is Redwood's theory. He was expounding it full force when I saw him at the precinct. He insisted on the cops getting hold of me, which they did out here, and I got back into the city as fast as I could—or as fast as *they* could. I came in a cop car with the siren going all the way. It's aged me ten years."

I couldn't help smiling at his obvious discomfort.

"Not that I wouldn't have aged anyhow," he went on, "with all the mess this has turned into. First the French manuscript doesn't turn up . . . then it does, with a dead body . . . then Redwood starts his KGB ruckus . . . the cops are swarming all over what was supposed to be a routine operation . . . tomorrow the papers will be all over it. . . ." He rubbed his stomach, a gesture with which I could fully sympathize. "You think you were glad to step off that plane; you don't know how glad I was to see you. This," he tapped the manuscript, "is the only thing that's gone right so far."

"Did the KGB do it?" I asked in alarm.

"That's Redwood's story. The city boys have their own ideas. And I'm reserving judgment."

"What do the police think happened?"

"Right now, I'd say, they're convincing themselves that she came home unexpectedly and surprised an intruder. That he hid in the bedroom and tried to get out while she was on the phone. Or maybe that he came in while she was on the phone. In either case, she hung up and saw him. She may have tried to struggle. The manuscript was in her hand. He bopped her with the candlestick and she went down. He panicked, grabbed the manuscript, and fled, leaving the door open. A little while later, Redwood shows up."

"The intruder must have heard Minnie talking excitedly about the manuscript," I said sadly. "I can just see her, waving it around in the air. That's how she was when she was excited. He must have thought it was very valuable."

"It's possible," Harrington said, getting up. "Obviously, then, you're willing to buy the intruder theory."

"Well . . . it could . . . I mean, it sounds reasonable."

"And the KGB theory?" He stood in front of me, and I felt the need to stand up and face him, though he was almost a foot taller.

"That too . . . sounds like it could be . . ."

"Reasonable? I have my doubts." He put a hand on my shoulder. "Look, I have those questions I still want to ask you, and I have to deliver you and the manuscript to the precinct—just temporarily—to demonstrate that you both exist. And Redwood will be happy to see you . . . or the manuscript, anyhow. So let's head back to the city. We can talk on the way in."

As he ushered me to the door, Dark Suit Two picked up the plastic-wrapped manuscript. Dark Suit Three opened the door and positioned himself as my escort.

"I'll take the manuscript," Harrington said, turning to Dark Suit Two. "You take Miller's bags."

Even in the CIA, rank hath its privileges.

We rode back to the city in Harrington's car, Dark Suits up front, one of them driving. They never spoke, and remained interchangeable and nameless to me.

"You've been helpful on one point already," Harrington said after a long silence. Either he had been composing his thoughts or simply waiting until we were well on our way.

"What's that?"

"The French manuscript. In a sense, we only have Redwood's word for it showing up. Though I suppose it's possible. And I have no reason to doubt Redwood. Besides, the city boys obviously accept it as real. I can see that, however, because it helps everything fall into place. If it's cuckoo . . . well, that's life. Lots of cuckoo things go on every day of the week."

"You don't believe it came into the office?"

"I didn't say that."

"Then how have I been helpful?"

"You didn't bat an eye when I told you that Kuzatov's French manuscript had shown up in the unsolicited pile. The fact—if it is a fact—makes no sense, but you don't bat an eye. Now, I can concoct various scenarios to account for your reaction—and I did, briefly—including a bit of collusion between you and Redwood, behind my back."

I started to protest, but he smiled and held up a hand.

"Forget it. I did almost immediately. But the point kept worrying me. As I said, it isn't worrying the boys at the precinct. And obviously it didn't worry you. That's what made me accept it—for the time being, anyhow.

"So, if I apologize for my narrative style, will you tell me what you can about 'over the transom' and how it's handled? It's important to me. And I do apologize for keeping you on tenterhooks."

"It's obvious," I said, "that you never worked in a publishing house, or you wouldn't be so hung up on this point. You must understand that publishing houses don't function like normal businesses."

"I'm beginning to understand that."

"First of all, there are two kinds of manuscripts, solicited and unsolicited. Solicited is everything that comes in from an agent, or"—I waved a casual hand at him—"from a known source. Unsolicited is everything that comes in unasked for or unexpected—'over the transom,' as it were."

"What are the proportions?"

"That's hard to say. It's different at different houses. We don't publish much fiction so that cuts down on the number of unsolicited manuscripts we get. I'd estimate that the average house publishes about one book in ten from the solicited pile and maybe one in a hundred from the slush."

"Slush?"

"The unsolicited manuscripts are put in what's called the 'slush pile.' I always called them 'slush,' but now 'over the transom' has become the popular term. It's more elegant."

"Now explain to me how they get read."

"Well, at Redwood we don't have 'first readers' the way larger houses do. Here the editors are their own first readers, and solicited material keeps them busy enough. The 'over the transom' stuff goes into a slush pile—and I mean pile. We're supposed to dip into it regularly, but . . . well, you saw what happened today." I shook my head in disbelief. "Today. All this happened just today."

"Some days are longer than others. It's been a long one for me too, Miller. Please . . . how do those manuscripts get read?"

"Well, sometimes we divvy up the pile, sometimes we dip into it on the honor system. Paul Ostrow more or less keeps track. About once every two weeks he asks if we've been reading our share, or reminds us to take a few manuscripts. Two weeks is usually the limit. They fill up the table after that."

"What table?"

"A table out in the bullpen. In the corner near Paul's office. When they come in, and they come in every day, the mail boy drops them off on the table."

"How does he know they're . . . slush?"

"He doesn't, really. Manuscripts from agents come in addressed to specific editors. Most of them are delivered by messengers anyhow, though some come in the mail. The thing is, if a manuscript comes in the mail and it's addressed just 'Redwood Press,' or 'Editor, Redwood Press,' it goes into the slush pile."

"Isn't it entered anywhere?"

"There used to be a log book. But we haven't kept one for a while."

"Why not?"

"Last year, in a fit of efficiency, Emory discontinued the practice. He thought it would speed up the reading if we didn't have to bother with the details. We don't like to hold onto manuscripts too long."

"Just a few weeks."

"Yes, but that's to be expected with material that we haven't asked for. A few weeks isn't long in this business. At least we read the slush. Some houses now are simply returning all unsolicited manuscripts unread."

"Do you really read these manuscripts?" Harrington's tone had that suspicious edge that all editors have come to recognize.

"You sound like every would-be author," I replied. "Yes, the unsolicited manuscripts get a fair shake. Maybe we pay more attention to the agented stuff, but that, presumably, has been screen-read by the agent. Let me put it this way: most of the slush is junk—plenty of agented manuscripts are junk—but a trained editor can spot junk in a very few pages. When it comes to the slush pile, an editor can go through maybe ten manuscripts in an afternoon."

"Skimming, of course."

"That may be, but skimming with a practiced eye. Look at it in terms of this situation—assuming Emory's story to be true. Minnie was able to spot Kuzatov's manuscript for what it was."

Harrington considered this for a minute.

"What you're telling me," he said, "is that the manuscript could have been on that table for anything up to two weeks."

"It's possible."

"And no one would have known."

He made it sound like a statement rather than a question, so I let it hang. We rode on in silence, and I stared out the window. We were on the Triboro Bridge and I could see the lights of Manhattan.

Once we were in Manhattan, working our way down the FDR Drive, Harrington came out of his torpor.

"Okay," he said, "I'll accept the manuscript business for now. Of course, there's checking to be done on the other end, Paris and elsewhere, but let's say I accept it. I still have a few questions, and I'd better get them in now, before we get involved with the circus at the precinct."

"Circus? Why do you say that?"

"When I left to meet you, the cops were getting ready to let the press in. Between a murder, which is always good for news, and the KGB theory that Redwood was cooking up, it'll be a circus."

"You don't really believe that, do you?" I asked. "I mean, you sound so . . . well, contemptuous."

He stared at me, as if weighing a new possibility. I turned away and resumed looking out the window.

"Damn!" he growled. "Listen, Miller, if Redwood cooked this up last week, and roped you into it, I'll have both your hides."

I turned to him angrily. "What are you suggesting?" I practically shouted. "Have you traveled ten thousand miles this past week on a crazy secret mission? Did you have to cart around a hot manuscript, wondering when you were going to be kidnapped or whether your plane was going to be blown up? Did you come home to find out one of your friends has been murdered? What makes you so sure the KGB isn't behind poor Minnie's murder? How do you know they're not following us even now?"

That stunned him. As a matter of fact, it stunned me a little, too. I hadn't realized how much I was bottling up.

Harrington eyed me warily. "Okay," he said. "Okay." He nodded his head slightly, not so much in agreement as in recognition of the fact that he was dealing with a hysteric.

"Let's take it a step at a time," he said, still nodding slightly. "Minnie Heffernan has been murdered, presumably while holding an important manuscript and maybe even waving it around in her hands. If so, because the manuscript isn't with her body, then the manuscript has been removed by the murderer. If so, then Redwood isn't the murderer, because he's still with the body when the police arrive. Or has Redwood managed to hide the manuscript somewhere?"

"Oh, come on now."

"I'm just ticking off possibilities."

"I thought you were trying to calm me down and convince me there's no KGB."

He breathed deep and let it out slowly. "That too, Miller. That too. Getting nasty may make you feel better now, but it doesn't help either of us in the long run. May I continue?"

I ignored him, sullenly, and he went on.

"So the murderer has removed the manuscript, possibly in panic, possibly in error or by accident, and possibly for reasons of his own. If the murderer is an intruder, then the likelihood is panic, error, or accident. As I say, the precinct boys seem inclined to follow up on that one. If the murderer is, as Redwood would have us believe, a KGB agent, then he may very well have removed the manuscript for reasons of his own. But how did he know the manuscript was in Heffernan's apartment? Nobody else did. And why did he go for the French manuscript? Or do you still believe he's after the Russian manuscript, possibly in one of those cars behind us?"

I resisted turning around to look out the back window.

"You've had a long week, Miller, and a trying one. Not that you haven't been splendid. As a matter of fact, you've done a fine job. I should have said that right at the start. I apologize for not thanking you properly. But then, I've been thrown off by . . . unexpected developments.

"Anyhow, it's been a hard week for you. And besides, you're an editor, you read a lot. I assume you have an active imagination."

"What's that supposed to mean?" I interrupted.

"Tell me, Miller, aside from imagination—which you're perfectly entitled to exercise, within bounds—do you have one shred of evidence that the KGB was onto you in Istanbul? Or anyone, for that matter?"

I shrugged. "How would I know?"

"Come on, Miller. Everything else in this business seems to have gotten screwed up, but the one thing that came off right is your part of it. We had a plan, a good one, and it worked—on your end, at least. I put it to you, aside from . . . nerves . . . you have no reason to believe anyone was after you or the manuscript, either in Istanbul or here."

"What about Minnie?"

"It could have been an intruder. I don't really know."

"But not the KGB?"

"It doesn't make sense. As if that counts for much in this mess."

"If it's not the KGB, then it has to be an intruder."

Instead of answering, he glanced out the window. "We're almost there," he said. "We'll have to get together again. I'll call you. But one quick question before we go in. Who else in the office, besides you, Redwood, and Ostrow, knows about the two manuscripts?"

"Fran Bishop, I would assume."

"Right. Redwood's secretary. But aside from her, who else?"

"No one. Emory made a whole point of playing up his trip to Paris, while I was just going to Istanbul on vacation."

"You're sure of that? I mean, no one dropped little hints or made some little slip?"

"Not to my knowledge."

The car slid to a stop and Harrington reached for the door handle.

"What are you getting at?" I asked his back.

"I just wanted to make sure."

"Sure of what?"

"That you, Redwood, Ostrow, and Bishop were the only ones who knew about the two manuscripts. Unless, of course, one of you spilled the beans to someone outside the office."

He stepped out of the car, and I was glad he couldn't see the expression on my face.

It *was* a circus at the precinct house, just as Harrington had feared. But we didn't get involved in that right away. Harrington went up to the desk, where an officer nodded his head wordlessly toward a hall to the left. Instantly, Harrington was off in that direction, and I hurried after him. I caught up with him in the doorway of a small office, into which he politely but firmly propelled me.

"Miller, this is Lieutenant Preston."

I smiled wanly at the large man seated at the desk. He raised a container of coffee in a vague greeting and finished chewing something. Sandwich wrappings and a brown paper bag littered his desk. While we waited for him to swallow some more coffee, Harrington closed the door and leaned his back against it.

"You certainly took your time," Preston said at last.

"My boys don't drive with a siren and lights, and they keep all four wheels on the road."

"Very civic-minded." He shifted his eyes to me. "You're Howard Miller?" Before I could answer, he added, "Wait." He lowered his eyes to the littered desktop and mulled something over in his mind, tapping the empty container against the sandwich wrappings. Finally he looked up at Harrington.

"We should do the Miranda, but if you're going to pull some fancy secrecy act on me or something . . ."

Harrington frowned. "I haven't crowded you once today, Preston. I've been remarkably cooperative . . . considering. The murder is all yours, and you know it. I'm here because of the manuscripts. You do what you have to."

Preston considered this. "I'd rather be off the record for now. Besides, we're not holding Mr. Miller. He'll be asked for a statement, of course, probably tomorrow in his office. For now, I just want to know one or two things. Just to tidy up loose ends."

"Fine," Harrington answered. "I'm sure Miller appreciates your respect for his rights. If it makes any difference, I'm a lawyer."

"His lawyer?" Preston waved his hand. "Never mind." He leaned back in his chair, looking very tired, and yawned and rubbed his eyes as he asked me questions.

"You're Howard Miller?"

"Yes."

"You're an editor at Redwood Press?"

"Yes."

"For the last five, six days, you've been . . . ?"

"In Istanbul."

"Were you there to pick up a manuscript?"

"Yes."

"Did you?"

"Yes."

He sat up straight and began to look more alert. "Did you have any trouble getting it?"

I wanted to see Harrington's expression, but figured I'd better not look at him. "Not really. No. No trouble."

Preston shifted his gaze to the package in Harrington's hand. "I don't want any details," he said, "but you had a plan for picking up this manuscript. The plan worked?"

"Yes."

"That's the manuscript?" he asked Harrington.

"Would you like to look at it?" Harrington responded.

"You know damn well I would."

Harrington handed it over, and we watched silently as Preston removed the pages from the plastic bag and examined them.

"I take it that's Russian."

"You take it rightly," Harrington said, retrieving the manuscript and returning it to the plastic bag.

"So that part of the story holds up," Preston said. "You must be very happy to have at least one manuscript. Even if the KGB has the other."

"Oh," Harrington replied evenly, "now we're buying that one? Things have changed since I was here a few hours ago."

"We're not buying anything. I was just reminded of your friend's rather novel theory."

"Where is he?"

"Doctor Strangelove? Oh, he's upstairs in the press room, explaining international intrigue to the fourth estate."

Harrington closed his eyes and breathed deeper than ever. His sigh was almost painful. "How much has he said?"

"To us, more than we care to know." Preston smiled maliciously. "To the press, God only knows. He's been up there with them for twenty minutes. I preferred to stay down here and enjoy a gourmet dinner."

"I cheerfully hope you choke on it, Preston. Did you have to give him a speaker's platform?"

"Shove it, pal. You know as well as I do that we couldn't stop him. What's the difference at this point?"

"Do you know what 'clandestine' means?" Harrington snapped. "Do you know what goes into an undercover operation? Or care? What difference? At any point in an operation, you try to rescue it."

"Well, lots of luck with rescuing this one. Short of strangling him, you're not going to get Redwood to shut up."

Harrington was edgy. "Are you done with Miller?"

"Yeah, I'm done with Miller—for now. Thank you, Mr. Miller, for your help. One of my men will come around to your office on a more formal basis, for a statement."

I started to reply, but Harrington grabbed me and yanked me out of the room. As we were leaving, Preston called out, "If you strangle Redwood, make sure you do it off the premises."

The press room, on the second floor, was big, grimy, barely furnished, and packed with people. Some were police in uniform, some were detectives, but most were obviously reporters and cameramen. Emory was center stage, so to speak, holding forth in his best Viennese-dentist manner.

". . . by seizing the manuscript," he was saying, "the KGB could effectively suppress its publication. That is why I was careful to broadcast information concerning only the one manuscript. All attention was directed to the Paris manuscript. Meanwhile, the true manuscript, in Kuzatov's original Russian, was safely delivered to my associate, who picked it up in Istanbul."

There was a flurry of questioning at this point, but Emory, having caught sight of us, ignored this. He raised his hands for silence. "And here," he announced, "is my associate now. Gentlemen, I give you a true hero, Mr. Howard Miller, who has saved the day for Redwood Press and Western democracy."

That was a bit much, but it served to distract everyone. The reporters and cameramen began looking around the room. A hand-held light picked me out. I could hear everyone bustling and shuffling and shooshing as I blinked into the blinding light. Somehow I moved forward to where Emory was waiting for me with open arms.

"Howie, dear boy," he greeted me, "how good to see you." He actually embraced me, as cameras clicked and strobe lights flashed and the TV cameras took it all in. With one arm around me, he faced me into the crowd and the damned lights. While the reporters shouted questions, my eyeballs began to function again.

"Where's the manuscript?" was the gist of the questions.

"In safe hands," I managed to mumble.

"Another of my associates has it," Emory said.

Harrington was across the room, near the doorway, and I tried to read his expression. But the lights and questions were distracting. Besides, Emory was speaking again, as if to deflect any attention from the tall man near the doorway.

"The important thing is that my plan to thwart the KGB has been successful," he stated. "This young man"—he patted my shoulder—"has assisted me in thwarting them. My very valuable associate." He continued to pat my shoulder as the picture-taking resumed, and I thought he was going to start on my head. But, ever the showman, he grabbed me again and switched to a serious tone.

"But at such a cost," he said with a mournful sigh. "Unfortunately, our very cleverness has claimed an innocent victim. The trick by which we outwitted the KGB has cost us dearly . . . another associate . . ." His sigh was mournful as he turned to me. "You have heard, Howard, about our poor, dear friend, Minnie?" He turned away even as I nodded. "This wonderful, wonderful woman," he proclaimed to the room, "so charming, so loved by us all—an innocent victim of those Russian hoodlums. I will not rest until they are apprehended."

There was a lot more in the same vein. Emory alternated between extolling my heroism and memorializing Minnie's charm, all the while proclaiming his own cleverness and the deviousness of the hateful KGB. It went over big with the press, though I could see Harrington, across the room, red-faced and glaring. As for me, I wondered what Emory expected me to make of his adulation. All that praise might have been welcome, if it were not uttered in the same breath with his fond recollections of Minnie. Did he really expect to convince me of his gratitude—which, I suspect, was genuine at the moment—while swallowing all that crap about dear, wonderful Minnie who charmed him so?

I managed to answer some questions, to spell my name for the reporters, and to avoid displaying the manuscript, which everyone wanted me to hold up for a picture. "Another one of my associates has it," Emory kept saying. I think by the third or fourth time he repeated it, Harrington had had enough of being another of Emory's associates. At any rate, sometime before I managed to break away from the press, I noticed that the CIA man was no longer in the room.

Emory, of course, monopolized the interview, so, when I finally excused myself, it was still going on. I staggered downstairs to the squad room, where a cop pointed to my valise and flight bag.

"Your friends left these for you," he said casually, and proceeded to find some paperwork to occupy him.

I picked up my bags, muttered, "Circus, indeed," and wandered out into the street. Finding a cab was a bit more difficult than in Istanbul, but I finally managed to make it home.

8

It was in the papers the next day, and on the morning telecasts. I never watch television in the morning, but I caught some of it on my all-news radio station while shaving. The *Times* actually put it on page one:

BOOK EDITOR SLAIN
IN HER APARTMENT
———
PUBLISHER FINDS BODY

The *News* tucked it away inside, but gave it a more dramatic focus:

PUBLISHER HELD FOR QUESTIONING
IN MYSTERY MANUSCRIPT MURDER
———
"THEY WERE AFTER THE BOOK" REDWOOD SAYS

By the time the *Post* splashed it across page one, it had assumed the shape that Emory was fashioning for it:

Redwood Charges:
"THE KGB
DID IT!"

CIA-linked Publisher Says
Russian Agents Killed Editor
for Dissident's Secret Book

Fortunately, I was out of the headlines, but that's the best I can say. Well, they spelled my name right. I can say that.

The stories were pretty accurate, as far as I could make out. The KGB angle was played up, not surprisingly, in light of Emory's long session with the reporters; the Paris mixup was played down, more Emory; and the business of the slush pile, too arcane for non-publishing intellects, was simply glossed over. My "heroic" role was muted in the *Times*, muddled in the *News*, and made much of in the *Post*. This simply meant that the ribbing I had to endure in the office reached its peak in the afternoon, after the *Post* hit the stands. Emory's CIA connection was barely hinted at in the two morning papers, and even the *Post*, which made as much of it as possible, managed not to mention Harrington. Poor Minnie, who should have been the center of attention, was crowded out by Emory. All three papers dutifully reported that the police were looking for an intruder.

I arrived in the office at my usual just-after-nine, nodded quickly to the receptionist, and skipped down to the partitioned-off hallway. Milt Foster was on the phone, facing out the window, as I passed his office. Myra Palmer's office was unoccupied, but Fred Snap was at his desk, and I had to nod a greeting.

"The traveler returns," he said, glancing over his granny glasses.

I managed a wan smile.

"Or should I say 'the conquering hero'?"

"Please don't."

"It's not every day you get your name in the paper."

"Nor do I want to."

"True," he remarked offhandedly. "Minnie got her name in the paper. And a lot of good that'll do her now."

I ignored that and continued on to my office.

There was a week's mail piled on my desk. Well, at least I would have something to occupy my attention. I really didn't feel like working, and dreaded the prospect of getting through what promised to be a long day. The phone rang. I let it ring three times while I hung up my jacket.

It was a reporter, the first of many. That whole day was punctuated by phone calls from reporters, friends, and passing acquaintances in and out of publishing. There was no five-minute period, while I was at my desk, that my thoughts or conversation weren't interrupted by the jangle of the phone bell.

The friends were solicitous, even admiring; the acquaintances were jocular; the reporters were a nuisance. The one call I would have welcomed, from Dinah, didn't come. Everyone asked me about myself; Minnie, if she was mentioned, was an afterthought. It made me feel guilty. The more I was reminded of my role in the "intrigue," the more I felt somehow responsible for Minnie's death. It was ridiculous, I told myself. Then I would recall Istanbul, and my "heroism" was even more ridiculous. It was a no-win situation, and I dreaded answering the phone. I compromised by leaving my desk whenever I could.

At least the people in the office spoke about Minnie. They couldn't resist ribbing me a little, but the murder was topic number one.

Milt Foster came in as I hung up on the first reporter.

"So, you sly devil, that was some vacation."

"Hello, Milt."

"Hello yourself." He wagged a finger at me. "Here you and I started this thing together, and you never even let me in on what you were up to."

"Emory would have had my head if I said anything."

"Oh, I know. I'm just kidding." He shook his head. "But who feels like kidding? How about that Minnie business? God!"

"Yes, Minnie."

"Do you think the KGB did it?"

"How should I know, Milt? I wasn't even here when it happened."

"But you were . . . part of the . . . I mean . . ."

"Don't believe everything you read in the papers, Milt. You, of all people, should know that."

"What does the CIA think?"

"I haven't the faintest idea."

"That's right." He nodded. "Better you should keep buttoned up."

"Milt, I honestly don't know what they think. But I did get to meet Myles Harrington."

"Hey, yeah!"

At this point the telephone interrupted us. Milt left while I answered the call. Before I was through, Paul Ostrow had entered my office.

"Welcome home," he said as soon as I hung up. "It looks like it's going to be a busy day for you."

"I guess the phone is going to keep up like this all day."

"The price of fame. Listen, I won't keep you. There're just two things I wanted to mention. The police will be in this afternoon for formal statements, so stick around. And we've been in touch with Minnie's sister about funeral arrangements. There'll be a service, but I don't know when or where. I'll let everyone know as soon as I find out."

"Paul, do you think the KGB killed her?"

"That's what Emory seems to think."

"Yes, but I'm asking you."

"Why should Emory make up a thing like that?"

"Well, for one, Myles Harrington doesn't seem to . . ."

The phone began ringing again.

"You'd better get that," Paul said, and ducked out.

A little later, to get away from the phone, I wandered down the hall. Myra was at her desk, listlessly thumbing through a manuscript. She looked pale and uncharacteristically subdued. When she noticed me in her doorway, she put down the page she was holding and began rubbing her face. Her eyes were teary.

"I can't work," she said.

"Neither can I."

"I keep waiting for Minnie to pop in. She always stopped by sometime in the morning. Her 'invigorating chat' she called it."

"You'll miss her the most of all of us."

Myra covered her eyes and shook her head. "She didn't belong here. I kept telling her it was a mistake to stay. A mistake." She locked her hands together and glared at me, or at the world in general. "A mistake. And now a damn fool mistake . . . a typical

Redwood Press foulup . . . has taken her. Poor, dear Minnie. She was always the innocent victim of that overbearing bastard."

"Emory didn't kill her."

"That's what I mean. It was his stupid manuscript. His big-deal manuscript that he couldn't even keep track of. And just because Minnie accidentally happened to have it . . . she was his innocent victim one more time. One last time. Oh, God. I hate that man."

"You should have seen him last night."

"I'm glad I didn't. I read about him in the paper, and I wanted to vomit. You certainly came out of this a big hero, according to the paper. That's how it goes. Some people come out covered with roses and others come out . . . dead. And the rest of us . . ." She pressed her hands against her forehead. "Just confused . . . terribly, terribly confused."

Finally, about noon, the call I was hoping for came in, but it didn't work out the way I wanted.

"My hero," she said as soon as I answered.

"Cut it out, Dinah. That's all I've been getting all morning long."

"But you deserve it, sweetie. Have you seen the *Post*?"

"Not yet."

"Go out and buy ten copies, for all your relatives."

"I don't have ten relatives, or ten who can read."

"Your picture's in it. They can read that."

"What does it say?"

"It says the KGB killed that lady in your office; that your obnoxious boss is linked to the CIA, as if we didn't know; and that you risked your precious, lovely neck to bring a top-secret manuscript back from Istanbul."

"Does it really describe my neck as precious and lovely?"

"There's a big picture—of the back of your head. The caption is: 'Close-up view of precious, lovely neck of hero. Neck is said to be adored by many females in publishing.'"

"Save me a copy. Or do you need it to wrap the garbage?"

"Sweetie, you sound bitter."

"Just a little out of sorts."

"Well, cheer up and tell me all about your adventures. But

just the highlights, because I have a lunch in a few minutes."

"Oh, of course, a *lunch*. Should I give you some tidbits that weren't in the papers? That way you could really spread some hot gossip."

"Howard, what's the matter?"

"I forgot that you're only serious about business. You might have called earlier. You might have asked me how I am. You might even have faked it and said you missed me."

There was a short silence and then she said, "I'll call you later. I have to run now. 'Bye." She hung up, and I stared at the phone, wondering what had gotten into me.

After a while, I went downstairs and bought a sandwich and a Coke and a copy of the *Post*. I ate at my desk, digesting my lunch and the newspaper article with equal dyspepsia. The picture was of me, all right, standing sheepishly next to Emory, weighed down by his arm around my shoulder. The article made more of Emory's CIA connection than he or Harrington must have liked, but it pretty much followed Emory's KGB scenario. My trip to Istanbul, without any substantial details, somehow came out as a romantic spy adventure.

That was good for more phone calls in the afternoon. I was distracted by them enough to be only dimly aware of all the hubbub in the office. The police were in, and afterwards I found out that their presence had been rather pervasive. They questioned everyone, took statements and, to everyone's distress, fingerprints, and spent more than an hour going over the slush-pile procedure. A detective I did not recognize from the previous night's encounter came in to question me late in the afternoon. He introduced himself and sat down.

"Well," he said, "you're the last, thank goodness."

"Saving the best for last?" I asked nervously.

"Hell, no. You're the only one we don't have to worry about. Unless you have some way of being in two places at once, we know for sure where you were. No, we left you for last because Lieutenant Preston has been closeted with your friend Harrington all afternoon and said to leave you in case he phoned in with any new question that might come up."

"And has any . . . new question come up?"

"Not that I know of."

"So what can I do for you?"

"Well, we need a statement, and your fingerprints. Though that's a formality. It helps us, though, so I'd appreciate it. You see, if we fingerprint *everybody*, including you, it makes all the other people less uncomfortable. And uncomfortable people tend to make trouble."

"Is this known as softening up the suspect? Lulling me into being unsuspecting?"

He laughed. "Mister, if your fingerprints match anything in that apartment, Preston is going to need a Ouija board to solve this one."

I laughed politely, but noticed that I was beginning to sweat. It was, no doubt, my having read *Crime and Punishment* at an impressionable age. The Raskolnikov syndrome. I would have to ask Dobbs about it.

The detective called in an associate, who handled the business of the fingerprints. I noticed that he made a point of letting people know that I was being fingerprinted.

"Now, about the statement," the detective said as I was wiping my hands. "Lieutenant Preston said I was to let you know we only want to cover your time in Istanbul and your flight home. No specific details, if you have instructions from Harrington. Just times and places."

"I have no instructions from Harrington."

"We just want you to know it's all right."

"I have no instructions."

"That's okay by me."

"I never even met Harrington until last night."

"You got your job, and I got mine."

"My job is editing books."

"It's okay by me. Just give us the times and places. Cover it all and be as specific about times as you can. Lieutenant Preston appreciates whatever cooperation you and the CIA can give us on this."

I was getting upset. I could see a picture forming of Howard Miller the CIA spy. People who had known me for years would cross the street to avoid me. Slowly, I wrote out an itinerary of my last six days, trying to calm myself by concentrating on fixing precise times to my actions. It was hopeless. I couldn't remember precise times. I didn't even know what time the plane landed. I put down the scheduled time and hoped for the best. I guessed at how long I had been at the airport with Harrington. What difference did it make, anyhow?

It took about ten minutes to write out the statement. I handed it over, and he read it. Then he nodded, asked me to sign it, which I did, and with a hearty handshake he was out of my office. I could picture him going down in the elevator, telling his associate, "That's the first time I ever took a statement from a spy. Notice how it doesn't say anything."

Now it was just four o'clock. Good God! Another hour to go. How was I going to manage? The phone rang.

"Hello."

"Howard Miller, please."

"Speaking."

"Speaking tersely, it seems."

"Who's this?"

"I just wanted to make sure I still have an editor, dear boy, and all in one piece."

I laughed and relaxed. "Yes, you still have an editor. I don't know about the one piece. Between the telephone and the police . . . and then there's just so much nonsense about Istanbul. . . . Tell me, when 'Crime Cabinet' became a big success, did you feel as if you wished you were back in obscurity again?"

Dobbs considered this for a moment and responded in measured tones. "I have never precisely known obscurity. And as for responding to instant notoriety, I have, I suspect, inner resources to call upon that apparently are not available to you. Does that properly put you in your place?"

"Quite properly. And it's good to talk to you. God, that's the first time I've said that today."

"Good. It suggests that you'd be delighted to talk to me endlessly. I've called to give you the opportunity."

"Now? On the phone?"

"Don't be ridiculous. This is *my* nickel, as you Americans say—an expression woefully outdated by inflation, but perpetuated by the phone company to encourage long conversations. I will make this one brief. Are you free for the weekend?"

"I am."

"Good. Are you up to spending it in New Jersey? I realize that for a New Yorker that involves considerable sacrifice."

"I'm up to it. At least I think I am."

"Splendid. Come out Friday direct from work, and stay till Monday morning. What with commuting between New York and Istanbul, the change will do you good. And I'm anxious to hear about your adventure."

"How do I get out there?"

"Take the one-sixty-nine bus to Kinderkamack, from the Port Authority. Enter on the Eighth Avenue side and walk past the newsstand, past the assorted petitioners who will be exhorting you to stop abortion and free Ireland, not necessarily in that order." I laughed and he continued, "When you see the dashiki-clad black man selling *The Bilalian News* and the beady-eyed white man holding a King James Bible and warning you that you will go to hell if you are a hippie, a homosexual, or a reader of any other version of holy writ, make a left. Ascend the stairs—the escalator probably won't be working—to the suburban bus level. Go to platform twenty-two. There are five different bus lines leaving from that platform and three of them are numbered one sixty-nine. Yours leaves from the middle."

He explained how to find his house and I hung up smiling. The thought of spending a weekend out of the city lightened my mood—until the phone rang again. This time it was Harrington. He came right to the point:

"Miller, I want to see you tomorrow. Can you meet me at one o'clock, or does that interfere with one of your publishing lunches?"

"I've given up publishing lunches."

"Don't let Redwood hear you say that. Fine, then, one o'clock. Meet me in the American Wing at the Metropolitan Museum. I'll be in the conservatory, in front of the bank facade."

"Where?"

"Don't worry about it, Miller. You'll find it. It's as big as the side of a barn. You can't miss it."

"One o'clock, Metropolitan Museum, American Wing, conservatory, in front of bank facade," I repeated, noting it down.

"Terrific. I'll see you then."

I hung up in a state of expectancy, which surprised me. I was actually looking forward to meeting Harrington at the museum. An assignation—the one thing I had dreaded in Istanbul. Well, one of the several things I had dreaded. But, of course, this was different. Harrington was on our side. He was safe. The museum was safe. But suddenly romantic. How clever of him to pick such an out-of-the-way place, at once public and private. I tried to envision the meeting. My mind was racing in different directions. Perhaps I should buy a trench coat after all.

"Woolgathering? Or perhaps reflecting on your new-found glory?"

It was Emory. He stood in the doorway, smiling benignly.

"Sorry, I didn't hear you."

"Ah, yes. Woolgathering. And when some of us have been so busy."

"Just taking a breather, Emory."

"Well, come breathe in my office. The air is better there."

Speaking of assignations, here was an unavoidable one, though I had been doing my best to avoid it all day. Actually, between the police being in and Emory being out much of the day, I had had no great difficulty in avoiding him. But I knew he would want to see me sometime. We had not exchanged a word since the night before, at the police station. I was due for a debriefing, or an interrogation, or whatever grilling Emory had in mind. Oh, well, it had been a lousy day; what more appropriate way to top it off? I followed him into his office.

"Close the door."

I remembered the last time he'd said that. It seemed like years ago.

"I'm very proud of you," he said, settling himself behind his desk. "It couldn't have gone better, even though I did plan it myself. But you handled everything exactly as I knew you would. It was a brilliant choice to send you."

This was hardly what I'd expected. I mumbled a thank-you as

I shut the door. But I was still wary as I crossed the room and sat down in the chair facing him.

"I really mean it," he went on. "I may even have meant some of those idiotic things I said about you last night. After all, in an unfortunate comedy of errors, you, at least, played your part exactly as planned." He shook his head. "So many unforeseen complications. But I knew I could count on you."

I felt like reminding him of what Harrington had said about his wanting to fly to Istanbul. But the rare experience of basking in Emory's praise was too welcome to interrupt.

"Everything went well in Istanbul?"

"Well, the restaurant proprietor who took my package when I came in wasn't there to give it to me when I left. I had to find it for myself."

"Probably a precaution."

"It certainly threw me for a minute."

"But only a minute. Obviously you recovered. And at the airport? Were you thrown when I wasn't there to meet you?"

That was a tough one to answer. It was one thing for me to have told Harrington I knew who he was. But how was I going to say to Emory that I'd known who Harrington was before I met him?

"Well," I finally managed, "Harrington sort of took me in hand right away. He didn't let me get thrown."

Emory nodded, apparently satisfied. "I would have preferred to be there," he said abstractedly. "But, of course, Harrington would have been there too." I realized that Emory hadn't concentrated much on my answer, which was a relief. I wondered what Emory *was* concentrating on, but knew that I would find out soon enough if he wanted me to. After a moment's pause, he seemed to snap to attention.

"We outwitted the KGB so successfully there," he said, "it's a shame we couldn't have done so here."

"You mean Minnie."

"Yes, poor Minnie. Innocent victims mean nothing to those thugs."

"You're . . . convinced . . . I mean, that the KGB . . ."

"Do you doubt it?" he asked tonelessly.

"Well, it's just that . . . you know, Harrington . . . the police . . ."

"What has Harrington been saying to you?"

"Nothing specific." I was getting flustered. "He just seemed so annoyed at your . . . theory. And the police don't seem to be going along with it either."

"The police!" he snorted. "They're determined to blame Minnie's murder on some unspecified, but highly-convenient-for-them intruder."

"Why convenient?"

"There are hundreds of breakins all the time in New York City. There are dozens of unsolved murders that can, no doubt legitimately, be ascribed to these breakers and enterers. It becomes convenient, in the face of unexplainable circumstances, to lump this murder in with those dozens of others."

"Why isn't it just as convenient for them to blame the KGB?"

"There are too many political ramifications. For one thing, the FBI would be brought into the investigation. The government would be sitting on top of everything. The State Department would be interfering. Don't upset the Russians." He laughed. "Oh, yes! Much easier to blame it on an ordinary housebreaker. Perfectly acceptable in New York these days."

"Harrington seems as skeptical of their theory as he is of yours. I don't understand what he's getting at."

Emory leaned forward confidingly. "Howard, my boy, you are being naive. Consider Harrington's position, and you will understand immediately the attitude he has adopted. He has had undue publicity showered upon what he would have preferred to be a relatively secret operation. The more I point to the obvious, and logical, KGB involvement, the more the unwanted publicity continues. He can't stand that. On the other hand, the police are eager to shunt the investigation off into a convenient blind alley. There will never be a solution to the investigation. The case will remain open indefinitely. He doesn't want that, either."

"So he sneers at both of you."

"Precisely. I'm glad you understand these things so quickly. But then, you have the advantage of having me explain them to you. So many things in this world seem so complicated or confusing because there is no one to explain them properly. I'm sure you're aware of that."

"Oh, yes."

"Have I told you I'm moving up the pub date of the Kuzatov book?" Emory asked with a sudden smile.

"No. How do you mean 'up'?"

"Moving it forward. It was planned as a spring book, or late winter. But I'm having the translation crashed. And I've been arranging things with the printer. We can have books by September first."

"That's fast."

"It's phenomenal. But these things can be done if you apply the proper pressures."

"Why the rush?"

"The message of this book is so important, it shouldn't be delayed a moment longer than is necessary."

I let that one go by.

"Besides," Emory went on, "I am very close to concluding a six-figure paperback deal—very high six figures. The price is conditional on my allowing the paperback people an early publication date—six months after the hardcover edition, instead of the customary year. Moving up our pub date makes the paperback date even more attractive."

I nodded and resisted comment. Six months was a major concession, more than just "an early publication date." It sounded like a take-the-money-and-run deal, with Emory mitigating its effects a bit by publishing the hardbound book as early as possible to cash in on its unexpected publicity. Of course, if the publicity could be kept warm long enough, that would help the paperback. I was assuming that Emory had already promised the paperback house he would keep the publicity as hot as he could for as long as he could.

Once again, Emory changed the subject and went back to rehash the "logic" of his KGB theory. I listened attentively, nodding and agreeing whenever it seemed appropriate. What I was thinking, however, was that Emory's insistence on hollering "the KGB did it" was a neat way of keeping publicity for the book as hot as he could. He might have been convincing if he hadn't told me about pushing up the publication date of the book and the six-figure paperback deal.

After a while he ran out of steam, and since I managed to look completely convinced, I was dismissed. I wandered back to my office to the sounds of people leaving. It was five o'clock at last. I sat at my desk and shuffled papers, waiting for the rush to be over. The phone rang.

"Hello."

"Oh, good. You're still there."

"Hello, Dinah. Yes, I'm still here."

"Are you still mad at me, Howard?"

"No, I'm not mad at you, Dinah."

"You sound awful."

"Thank you. That's just the kind of cheering up I need."

"I called to apologize, Howard. I thought maybe by now you had gotten out of your self-pitying mood."

"Well, I haven't. And apologies aren't necessary."

"Listen, mister, if I call to apologize, you damn well better be prepared to accept any apologies I offer."

"That sounds more like you. Okay, apologize."

"I'm sorry I upset you when you were feeling sorry for yourself. I'm sorry I didn't ask how you were. How are you?"

"I'm miserable, thank you. Come on, the rest of it."

"The rest of what?"

"You're supposed to fake it and say you missed me."

"Don't press your luck, sweetie."

"I missed you."

"I bet you did. In between all the running around with CIA agents and KGB agents, you didn't even have time to send me a postcard. You might have sent me one lousy postcard."

"I thought of you. If the KGB got me, I was going to die with your name on my lips."

"Oh, fine! And then they'd come after me. You're a real pal."

"What are you doing tonight?"

"I'm cooking you dinner."

"Oh."

"See, it was easier than you thought. And then, after you help me with the dishes, I'm curling up on the sofa and listening to you tell me all about your adventures."

"Can I curl up on the sofa with you?"
"Only if you keep a sword between us."
"Forget it. What time should I be there?"
"Seven. Can you remember how to find it?"
"The address is engraved on my heart."
"That must look great on your EKG. See you at seven. 'Bye."
So it turned out to be not such a lousy day, after all.

9

How I got through Thursday morning I can't even remember. There were conversations in the office, I'm pretty sure; phone calls, of course; and even a certain amount of work. Emory was not around, which made things easier. At noon sharp, I ducked out, grabbed a sandwich and a Coke, and headed uptown for the Metropolitan Museum of Art.

There were signs at the museum directing me to the American Wing, and once I arrived, I knew what Harrington meant by the conservatory. A spacious, glass-roofed courtyard, it was decorated with foliage as well as art objects and featured a stunning view of Central Park. At the north end of this indoor-outdoor area was the facade of a bank that had originally been erected on Wall Street in the 1820s.

Harrington was standing facing the bank facade, examining its marble columns and elegantly framed windows. I came up and stood next to him, fixing my gaze on the columns, which are at second-floor level.

"I hope I'm not late," I said, not turning my head. "Have you been waiting long?"

"No. And you're right on time."

"What a clever place to pick for a meeting."

"Miller." He turned his head to look at me. "I appreciate that you have seen countless spy movies. If you want to keep staring at those marble columns, that's okay with me. But will you please move your lips when you talk. It's very difficult to understand you."

"Oh, sorry," I mumbled.

"And I picked this place because I wanted to see it, and my time in New York is very limited. I've already been through the wing, but if you want to see the exhibits, I don't mind going through again. We can talk as we walk. And then, when we're through here, I want to go over and see the Temple of Dendur."

"Sure," I replied, disappointed by his casual attitude.

"Okay, but let's turn around first and look at the stuff here. There's some lovely Tiffany things and a Louis Sullivan staircase."

"Oh, good," I said, relaxing a little. "I was just in Chicago recently and saw Louis Sullivan things there."

It turned into a pleasant hour or so, despite Harrington's constant probing. He had me go over Istanbul minute by minute, and made me repeat some things. But I didn't mind. Between the informality of the grilling and the pleasure of some of the paintings, I was growing steadily more relaxed. We were standing in front of the enormous painting of Washington crossing the Delaware when I told him about the taxi ride back to the hotel after the restaurant, and I began to laugh about opening the window to dispel any gas. He didn't laugh with me, but then he didn't criticize or make any remarks about my seeing too many spy movies. In general, he remained expressionless, just constantly asking questions.

We were heading out of the American Wing, in search of the Egyptian section, when he asked me to go over the restaurant pickup again.

"I'll do it, if that's what you want. But I think I see what you're getting at."

"What's that, Miller?"

"You're making me go over all this in a more relaxed mood . . . in a more objective way . . . so I can see that the KGB wasn't really after me."

"And can you see that?"

"I think so. And there's something else."

"What's that?"

"I think Emory was fostering my . . . I don't know . . . my fears, or my vivid imagination. He was setting me up. Just as he's trying to do now, to convince me that it's logical that the KGB killed Minnie."

Harrington considered this. "I don't know if I'd put it that way," he said finally. "But he certainly preferred to have you go to Istanbul instead of Ostrow. From the time you first intruded into this affair, he had a preference for sending you."

"Why?"

"I don't know. At least I'm not sure. He may simply have preferred to have Ostrow on hand in the office. I'd prefer not to speculate on it at this point."

We arrived at the Temple of Dendur with the conversation having run down, so we examined it silently. We both seemed to prefer this, and I found the small Egyptian temple, in its impressive setting, rather moving. As I turned away from it, after having walked all around it and lingered for a second overall look, I noticed that Harrington was seated on a nearby bench. I walked over and sat down next to him.

"There's another point about Istanbul," he said.

I gazed at the temple and waited for him to continue.

"If the KGB wasn't after you in Istanbul, why would they have gone after Minnie . . . or the French manuscript, which was as good as lost anyhow?"

"You really don't buy Emory's KGB theory."

"I wouldn't pay much for it."

"How much would you pay for the police's intruder theory?"

"That's a funny way of putting it. How much is it worth, you mean? How sound is it? I haven't had time to discuss the investigation with Preston. But I'll make time, you can be sure. Meanwhile, let me put it this way: the police are no dopes. If they suspect an intruder, and stick to that line, as they are sticking to it, then it's a safe bet they have a reason for doing so."

"Emory says they have a good reason."

"And what, pray tell, is that?"

"He says they're up a tree and it's convenient for them to blame it on an intruder, because breaking and entering is so common in New York."

"Full of theories, our friend Redwood."

"You sound unconvinced."

"Are you? Convinced, that is."

"I don't know. I think I'm just confused. But it stands to reason that it's one or the other. Either the KGB killed Minnie or an intruder did."

"You don't see a third possibility?"

"A third? No." I frowned at him, perplexed. "But you do, obviously."

"See a third possibility? Oh, yes. I'm sure that if you think about it, you will too. And maybe even a fourth or fifth. But seeing possibilities and proving actualities are different—worlds apart." Suddenly he stood up, saying, "Enough museum for today."

I remained seated, mulling over something that had begun nagging at my conscience.

"I'm going to see Hartley Dobbs this weekend. He's invited me to stay at his house in New Jersey."

"The TV person? That should be pleasant."

"He's an author of mine. We've become close. I kind of used him as a confidant before I went to Istanbul."

Harrington sat down again. "What are you getting at?"

"I told him about my trip . . . about the two manuscripts."

"You . . . *what?*" He slapped his head. "First it's that mixup with the manuscripts, then it's a murder, then it's Redwood's KGB nonsense in all the papers. Now you're telling me we're going to be dramatized on network television."

"I didn't say that."

He jumped up again and began pacing up and down. Slowly, visibly, he began to compose himself. "Sorry, sorry," he muttered. "You're quite right." He stood in front of me and smiled sheepishly. "Just a display of temper on my part. I do it from time to time to show I'm human."

"That's all right."

"I overreacted. Dobbs will be all right. Do you know him?"

"I told you, he's my author."

"I mean as . . . Do you know his background?"

"He was a screenwriter in Hollywood for several years before he did the television series."

Harrington stared at me as if considering something. "I mean before that." He shook his head. "Obviously you don't. Never mind. I'm sure we can rely on him."

"He's not going to turn this into one of his 'Crime Cabinet' scripts."

"It would be interesting to see what he makes of it if he did. As a matter of fact, it would be of interest to me to know what he thinks of all this. I mean, an expert on famous crimes and all."

"Are you suggesting that I discuss with him everything that's happened? A minute ago you were . . ."

"I know, I know." He held up a hand. "Don't mind me, Miller. I was excited before. I know Dobbs isn't going to make a TV show out of all this. But it would be very helpful for you to discuss everything with someone like Dobbs. It'll give you perspective. A man with his background will be expert at providing perspective." He laughed and slapped his leg. "It's settled then. You can talk to Dobbs and thrash it all out with him. I'm sure it'll be helpful."

"I think it's time to get back to the office," I said after a pause.

"Yes, go ahead. And thanks for meeting me up here. It's been pleasant. And educational. You go ahead. I'm going to sit here a minute and contemplate the temple. It's nice to imagine myself three thousand years away from things. That's how I get my perspective."

I smiled and waved goodbye. He nodded and was instantly lost in some private thoughts. On my way out of the museum, I tried to figure out his sudden switch about Dobbs. It seemed out of character for Harrington to be so agreeable about my talking over things with a television personality. But the whole interview remained a puzzle to me. Oh, well! Maybe Dobbs would be able to straighten it out on the weekend.

It was Friday morning and I could see that it was going to be a very long day. I had no lunch date and no appointment with Harrington. Emory had already favored me with a debriefing, or briefing, as it turned out, so I didn't even have that to look forward to.

The best cure for a long day is to have a manuscript to work on, but I was fresh out. There weren't even galleys from one of my books that had to be checked. The next best thing was to read a manuscript. Given the constant interruptions of an office day, most editors prefer to read manuscripts at home. But here I was with time on my hands and a pile of newly arrived manuscripts on the corner of my desk. I chose the likeliest-looking one and began to read. The first interruption came just as I started page three.

Sherwood Leitner posed nonchalantly in the doorway.

"I haven't seen you for days."

"I've been here," I replied, putting down the manuscript.

"You took a long lunch yesterday."

"And I stayed late to make up for it." Actually, I had been somewhat conscience-stricken at returning from the museum after three.

"Golly, fella, I'm not trying to keep tabs. I just thought maybe it had something to do with the investigation."

"Like what?"

"I mean . . . you know . . . like you're part of the . . . what's going on."

"Sherwood, if you mean was I out with my fellow CIA agents, chasing the KGB, the answer is no."

"You don't believe that KGB crap do you?"

"You don't believe I'm a CIA agent, do you?"

Sherwood lingered in the doorway, hesitating over whether to continue the conversation. "I think that's my phone," he said suddenly, and hurried off to his office.

Mr. Snap followed, almost immediately, only he didn't stop in the doorway. He nodded a greeting, accompanied it with a grunt, and came in and sat down in the visitor's chair.

"God, this is lumpy."

"I think Emory intended it to discourage visitors."

"I'm not a visitor. Try to think of me as your employer."

"I do. Whenever I try not to think of Emory as my employer."

"What's that supposed to mean?"

"It's my attempt at currying favor."

"Why are editors always wise guys?"

I put down the manuscript again, having gotten all the way to page four, and tried to look attentive.

Mr. Snap shifted in his seat and leaned forward. "What's Harrington like?" he asked.

For a second I thought he had followed me yesterday afternoon. Then I realized he was referring to last Tuesday night.

"You've never met him?"

"I wouldn't be asking if I'd met him."

"He's tall, at least six feet, and just a little beefy. He tries to look like a businessman . . . you know, conservative suits. Only when you look at him, you kind of know he's not a businessman."

"What's that supposed to mean? I'm a businessman. When you look at me, how do I look?"

"Like a businessman."

He mulled that over for a bit and decided to try another tack. "What's he like to talk to?"

"Very agreeable."

"Yeah? I've spoken with him on the phone once or twice. He always sounded as if his words were costing him."

"Oh, no," I said a bit eagerly. "He was quite talkative to me."

Mr. Snap peered at me over his granny glasses. "Maybe it's your lovable personality. Maybe he's trying to recruit you. Maybe he has already recruited you."

"Oh, for God's sake, not you too!"

"Yeah, I heard our friend Leitner before. Let me warn you, Miller, that half this office thinks you're connected with the CIA."

"Only half? What about the other half?"

"The other half doesn't think about anything."

"It was pure accident that Emory sent me to Istanbul. Didn't he tell you that? Didn't he tell you that he was originally going to send Paul? It was only because I stuck my nose into things that Emory picked me. I never even met Harrington until Tuesday night at the airport."

"You turned the manuscript over to him fast enough."

"He was waving a CIA identity card at me. He had a couple of hoods with him. He said Emory was in jail. What was I supposed to do?"

"Do you know where the manuscript is now?"

"The one I brought back? No."

"Emory picked it up on Wednesday. It's out for translation."

"Then, if you know, why did you ask me?"

"I wanted to see if you knew."

"Christ! You're as bad as the others. You think I'm a CIA agent too."

"Just checking. Tell me about Harrington."

"I've told you about Harrington. Ask Emory about him."

"Emory only tells me what he wants, or what he thinks I have to know for business reasons. You're his favorite."

"His favorite!" I practically shouted.

That ended that conversation. Four minutes and three pages later, it was Milt's turn.

"If you're coming in to check my CIA papers," I said, "just turn around and march right out again."

Milt laughed. "Yeah. Say, I'll bet half the people in this office think you're working for the CIA."

"So I've heard. Which half are you siding with?"

"Oh, I know you aren't CIA."

"How can you be so sure?"

"Nah! You would have really had to fool me . . . really fool me. And you're not that good."

"Gee, thanks a lot."

"No offense. What I meant was, you're too honest, too open. Whatever you're thinking, it's written all over you."

"I don't know," I muttered. "I don't think of myself that way."

"We never do," Milt replied breezily. "I think of myself as Cary Grant. Anyhow, I didn't come in to ask you if you're CIA. I was wondering about something else. Did the cops really fingerprint you?"

"Yes."

"I find that interesting. Don't you?"

"They explained it easily enough. They said if they fingerprint every single person in the place, even someone who wasn't around at the time of the murder, it makes for less grumbling. And they prefer it when there's less grumbling."

"And you bought that?"

"Why not?"

Milt paced up and down, apparently trying to compose his

argument. I waited patiently, reconsidering what I had said and satisfied with it.

"Look," Milt said, "the police obviously dusted Minnie's apartment for prints and picked up a load of them. Right?"

"I suppose so."

"Now most of them, of course, turn out to be Minnie's. But there have to be others. Minnie probably had a cleaning woman, for instance. But she also had visitors. Who would they be?"

"I don't know. Her friends, I guess."

"And who would her friends be?"

"I never met any of Minnie's friends."

"How about people in this office?"

"Don't look at me. I've never been in Minnie's apartment."

"Neither have I," Milt said. "And I'm sure Emory hasn't. He couldn't even stand going into her office. Maybe Myra has—they were friends, sort of—and maybe Paul or Sherwood; we could ask them. But that's not the point. If Myra's fingerprints showed up in Minnie's apartment, or even Paul's or Sherwood's, they could be explained."

"So what are you getting at?"

"Fingerprinting everyone here was just a blind. At best a necessary precaution."

"I don't follow you."

"Look, the apartment must be loaded with unidentifiable prints. Everyone's apartment probably is. A delivery boy . . . a mailman delivering a package . . . the plumber comes to fix the sink, or something."

"So what's your point?"

"Don't you see? The police go through a whole *megillah* of lining up matched prints. But they still wind up with some—even one—that can't be identified. *Voila!* The intruder. They're just horsing around to build up their case that a burglar came in and, when Minnie surprised him, conked her one."

"Then you don't believe it was a burglar?"

"Oh, it could have been. I'm just saying the police are making damned sure that's how it's going to work out."

"What about what Emory is saying?"

Milt snorted. "The KGB?" He gave me a supercilious smile.

"Don't tell me Emory has you brainwashed. Come on, we're all big boys here."

"It's just as logical as an unknown intruder."

"Maybe if someone other than Emory had suggested it, I'd be willing to consider it. What does your friend Harrington say?"

"Why does everyone think he's my friend?"

"What does he say?"

"I'm not at liberty to divulge that information."

"Bullshit! Oh, what an unconvincing liar you are. Now I'm sure of it. He thinks it's as big a crock as I do."

"Maybe."

"Maybe, my ass. Emory is just riding that horse for all it's worth in publicity. No fool he. Not our Emory. No! Anyone with an eye on this case has to know what he's up to." Milt rubbed his chin thoughtfully. "Maybe the cops are stacking the deck, maybe they're moving in a preconceived direction, but they're probably right. It probably was an intruder. I bet Harrington believes that too."

I decided not to correct him.

About thirty minutes later, I had gotten through less than that number of pages, a sign of my lack of concentration, when Myra came in. She flashed her big teeth in a wicked smile.

"Back to the humdrum after his glamorous spy career."

I put the manuscript down and glared at her.

"Don't tell me," she said, snapping those teeth. "You are not a spy. I rank that with those other notable statements: 'I am not a crook,' and 'I will never lie to you.' But I don't assume you'll be running for president."

"Do I look like presidential material?"

"You're a boot-licking fascist lackey. That should help."

"I take it you're referring to my vacation in Istanbul."

"Yes, vacation—that's a good one. Running errands for your CIA masters."

"Just a little errand for Emory, a small boon for Western civilization."

She stalked over to the chair and flopped into it. "My God, you're even more depraved than I thought. You actually believe you did something wonderful."

"Not wonderful. Just useful."

"Oh, you're hopeless." She leaned forward in the chair. "Listen, I came in here for a reason. Do you think you could put aside your James Bond fantasies long enough to discuss something closer to home than Istanbul?"

"I'll try."

"Good. Tell me . . . what are the police up to?"

"I have no idea. You probably know as much about it as I do."

"A likely statement. Look, I'm not trying to suggest that the police have taken you into their confidence. They may be twits, but they're also paranoid by nature. It's just that you've had more contact with them than any of us. I thought maybe you had an insight into the workings of their so-called minds."

"Myra, let's get something straight. Except for a brief conversation with Lieutenant Preston on Tuesday night, in which he said almost nothing, I've probably had less contact with the police than any of us. They're not interested in me. I wasn't even here when . . . when the murder took place."

"They questioned you on Wednesday. I saw them in your office."

"They took a statement . . . on my movements in Istanbul and on Tuesday, when I flew home. And they took my fingerprints."

"They took everyone's fingerprints."

"Yes. And some people apparently objected. Were you one of them?"

She bristled at this. "Certainly not! Why should I? I'll do anything to help find Minnie's murderer. If taking everyone's fingerprints can help, I was glad to do it."

"Okay, then. Maybe it will help. Though Milt Foster thinks it's just a dodge to help them build a case against an unknown intruder."

"Foster," she said contemptuously. "Another great brain. Though, in this case, he's probably right. They certainly seem to be fixed on the idea it was a burglar. Poor Minnie. Just another police statistic." Her face contorted and she seemed near tears. "Oh, I'd give anything to be able to prove who it was. I owe that much to Minnie's memory."

"That's very loyal of you."

"I don't need comments about loyalty," she snapped, recovering herself, "from the Benedict Arnold of the publishing business."

"That's not fair. Besides, it's inappropriate."

"That's true. Drawing and quartering would be more appropriate for anyone who helped that bastard in any of his schemes."

"I also helped Kuzatov."

"Naturally. All you CIA lackeys have to stick together. Are you going to persist, then, in the fiction that you know nothing about the police investigation?"

"My lips are sealed."

"Would that they remained that way forever." She stood up. "I'll have to content myself with small blessings. It was a mistake coming in here for an intelligent conversation, anyway. I should have known better." In the doorway she paused. "And when you report this to your masters, make sure you spell my name right. Or has it all been recorded on tape?"

I picked up my manuscript.

That left Paul, and he came in just before noon.

"Have anyone on for lunch?"

"Just my normal appetite."

"Want to try the new Chinese restaurant on Twentieth Street?"

"I'm always willing to try a new Chinese restaurant."

"Good. I'll pick you up in about ten minutes."

In the restaurant, we chatted amiably about the decor and the menu. Even after we ordered, Paul kept the conversation innocuous. We were mopping up the last of a passably good meal, and Paul was pouring tea, when he finally came to the point.

"Emory says that Harrington doesn't buy his KGB theory."

"Do you?" I parried.

"It's the only thing that makes sense. Well, no, let me put it this way—it makes more sense than the intruder theory."

Good old Paul, loyal to the core. I wondered fleetingly why Emory had changed his mind about sending someone so perfect to Istanbul. Maybe he really did suspect some danger when the French manuscript hadn't turned up. Maybe he didn't want to risk Paul, an old friend, a family man. Maybe I was just more expendable.

"You seem doubtful."

"What? Oh, no," I said hurriedly. "I was just mulling over the pros and cons. As I said to Harrington, it's got to be one or the other."

"And what did he say to that?"

"Well . . ." I began, pausing to sip some tea. I stared innocently, or so I hoped, into Paul's questioning eyes. When you stare into someone's eyes, I heard somewhere, he's supposed to believe you're not lying to him. "You can imagine that he's not happy with the KGB theory, and with all the newspaper and TV attention Emory is getting. But it's not just that. He said a couple of times that the police are no dopes, that they know what they're doing, and that if they think it's an intruder, they must have good reason to think so."

"When did he say this?"

"Yesterday." I continued staring at him over my teacup. "He called me in the office to check on a few things. We chatted briefly about the police investigation." My eyes were getting tired, and the tea was burning a hole down my insides. "And, of course, when I was with him Tuesday night, he was furious with Emory. Very mad. He called it a circus."

"A circus."

"Yes." I put down the cup and relaxed my gaze. He either believed me by now or not. I couldn't keep looking at him.

"It's all the publicity, isn't it? That's what's really bothering him."

"Oh, yes," I agreed. "Emory's quite right about that. Harrington doesn't *want* it to be the KGB, so he's convinced himself the police know what they're doing."

"Yes, Emory usually is right about things. And that's all that Harrington said about it?"

"That's all. He called to check about Istanbul. Mostly he had me go over the whole thing, step by step."

"Which you did."

"Sure. Why not?"

"No reason. None at all. We're all cooperating—even Emory. Even if the police and Harrington don't believe that." He shook his head sadly. "It would be so much better, though, if they understood that Emory is right, that he isn't just making it up."

Back in the office, I sat at my desk and stared into space, pondering why I had suddenly decided to lie. Not because it was Paul—that was really just lying to Emory. And not about the museum—that was a bit of self-protection. No. Why had I decided to keep mum about Harrington's hints about a third possibility? It had come over me in an instant, as Paul was asking about what Harrington had said, a tingling in the back of my neck and an instinctive response: *Lie.* I hoped I had pulled it off. But why had I done it? Oh, well, Dobbs would probably have an answer when I discussed it with him over the weekend.

So there were at least two things I was shutting out of my mind until I could speak with Dobbs. It was just as well, because the trip required my full attention. The bus ride was hot, stuffy, and punctuated by unnecessarily jolting stops and starts as we crawled through traffic. After a few minutes, my disposition was as sour as the expression on the driver's face. I worriedly sought out the few landmarks Dobbs had mentioned. The fact that the window was plastic made this difficult, because it was completely clouded over and scuffed.

But I arrived in Kinderkamack in one piece, although worn out. Some world traveler. The bus reached Elm Street precisely as late as Dobbs had predicted. "Pay no attention to the printed schedule," he had said. "It is published by the bus company as a form of sick humor." The houses were old and set back from the sidewalk about fifty feet. Number 21 was a two-story Cape Cod with green clapboard facing and white trim. Dobbs opened the door almost as soon as I rang the bell.

"Dear boy," he greeted me, grabbing my bag. "You made it. Would you like to lie down? Perhaps a cold compress?"

"It wasn't that bad," I said, entering the house.

"Nonsense. I've made the trip many times. I am well-acquainted with the pandemonium of that terminal in rush hour and the indignities of that bus ride." I followed him into a small hall and up a flight of stairs. "Omnibus of New Jersey has two criteria for employment: you have to be surly and stupid. If you are more surly than stupid, they make you a driver; if you are more stupid than surly they make you a dispatcher. If you are overwhelmingly surly *and* stupid, they make you an executive of the company." He waved me into a bedroom and deposited my bag. "This is the guest room. Last occupied, as I recall, by an aged actress whose heyday was in the reign of Edward the Seventh. Are you sure you don't want to lie down for a while?"

"No, I'm all right. But I think I'll change."

"Certainly, certainly. And by all means get into something informal. We don't dress for dinner here." He himself was in white ducks and sneakers, with an open shirt and a cravat loosely tucked in at the neck, looking like something out of the British Colony in California.

When I had changed into something informal—a plaid sport shirt, jeans, and loafers—I wandered downstairs. Dobbs was waiting for me in the book-lined living room. There was a large window in one wall, a fireplace in another, and some doorways, but every possible inch of wall space was lined with shelves, and every shelf was crammed with books. I began to inspect them, something I do almost automatically on first coming into a room.

"See if you can divine the system," he said, getting up from an armchair near the window. "Meanwhile, I'll prepare the drinks." He left, presumably for the kitchen, and I wandered slowly from shelf to shelf around the room. In a few minutes he was back, carrying a tray and smiling. On the tray were a bottle of cognac, a bottle of Perrier, an ice bucket, two glasses and some stirrers.

"I thought I'd join you in brandy and soda. That way, we'll both be ready to continue with the brandy after dinner." He put it all down and looked at me archly. "Figured it out yet?"

"No problem," I replied. "I keep mine the same way. American, British, and European literature arranged separately in roughly chronological order by author."

"It's really the simplest way."

"Sure. If you can remember that Dickens comes after Jane Austen and before Conrad."

"If you can't, you don't deserve a library."

The discussion stayed on books over drinks, and pretty soon I was not only relaxed, but mellow. We were seated in chairs near the window, and after a while the fading light reminded Dobbs that we must be getting hungry. He finished his drink, his third, and stood up.

"I'd better prepare dinner."

"I guess I am hungry," I agreed.

"Can you make it into the kitchen? I'm not that efficient that I can't use help. An old bachelor living alone in a house in the suburbs has to, perforce, become something of a cook. 'Something' is about the best I have attained. James Beard I'm not."

"I'm sure you'll do," I said, rising unsteadily. "I can't even boil water. If there was a restaurant strike in New York, I'd starve."

I did manage to set the table. While dinner was in preparation, I stood admiring his culinary skill and the pleasantness of his kitchen.

"What a neat setting," I said, plopping into a chair, "and such a clean tablecloth."

"Good heavens," he said, surprised. "What were you expecting?"

"I was just thinking of my famous assignation meal in the restaurant in Istanbul."

"Comparisons, they say, can be odious."

"That restaurant was certainly odious. The tablecloth was so dusty it was like eating in Miss Havisham's room in *Great Expectations*."

"That's called the delights of foreign travel," he said, bringing two heaping plates of chicken and rice to the table and sitting down. "If you don't mind, however, I would prefer to leave your famous assignation restaurant for its proper place in the saga. I want to hear everything, of course, but in sequence and from the beginning."

"Suits me," I said between mouthfuls. "Besides, I don't think it's good for our digestion to discuss Turkish cooking while we're eating."

"Is there anything about Istanbul that is palatable?"

"Oh, the Blue Mosque. That's not only palatable, it's delectable. And the Topkapi." We ate in silence while I gathered my thoughts. "You know," I said finally, "now that I'm safely thousands of miles away, there are aspects of Istanbul I can recall with pleasure . . . a modicum of pleasure."

He laughed, and I began chattering happily about the sights of Istanbul. I was still at it while we did the dishes and put them away. Before closing the cabinet, Dobbs removed two brandy snifters.

"It's time to repair to the sitting room, as they say, so you can a tale unfold. And remember, I want it all, every last detail, in proper sequence, from beginning to end."

Back in the living room, we drew the drapes, poured the first of several brandies, and settled ourselves in our chairs. And I began talking. It poured out of me—all that I had done, all that I thought, everything that had happened to me, from the afternoon that Emory mentioned trouble about the French manuscript to the point where I landed at Kennedy and heard my name being called on the loudspeaker. Dobbs stopped me from time to time, asking me to repeat something or to explain where places in and around Istanbul were in relation to each other.

The session went on for hours, until we had killed the bottle and I was practically falling on my face. I staggered up to bed and slept right through the rest of Saturday morning. In the afternoon, Dobbs took me for a walk through Kinderkamack, commenting all the while about small-town life. It was his contention that most towns in northern New Jersey were true American small towns and not what we think of as suburbs of a great city. They could just as well have been in Iowa, say, for all their geographical proximity to New York. We ate an early dinner in a local restaurant and spent the remaining hours of daylight sitting under the trees in his backyard.

When it grew dark, we returned to the living room, opened another bottle of brandy, settled in our chairs, and picked up my saga at the point where I first met Harrington. I covered it all: the airport, the police station, the office, everyone's comments, the session with Emory, the meeting with Harrington at the museum, lunch with Paul. At one point, Dobbs said I was skipping over something. I said no, I wasn't, but he persisted, and I realized that I had been leaving out Dinah. This amused him, and he made me backtrack to the party where I first met her, and to our night in Chicago. Throughout the whole recitation he would make me go back over things, asking questions that hadn't occurred to me, giving me a new focus on events or on people's reactions. Once again, I talked on into the wee hours, though thankfully we didn't kill another bottle.

I was able to get up at a reasonable hour on Sunday. Dobbs cooked us an enormous breakfast, and afterwards we spent hours in his backyard, working our way through two newspapers. He asked me if I wanted the *Times* crossword puzzle, but I declined and then watched enviously while he breezed through it. We ate a kind of potluck late lunch or early supper, and went back out to sit under the trees.

"So there are two things," he said, "that you have left for me to unravel. Two things that your overburdened mind refuses to cope with. One is Mr. Harrington's enigmatic hint, and the other is your even more enigmatic compulsion to lie to Paul—or rather, to withhold information from him. Are you sure there aren't other Gordian knots in your tangled recollection that need my Alexandrian sword?"

"There probably are, but they haven't occurred to me."

"Certain things do seem not to occur to you. But then, other things do. You appear to have seen through Redwood's motivation quickly enough. On the whole, I'd say your perspicacity is about average."

"Par for the course?"

"Some courses," he said, suddenly thoughtful, "should be played at better than par. I have a feeling that this is one of them. It strikes me that this is a landscape so pleasantly obvious that one should walk through it with his eyes close to the ground, looking for hidden hazards."

"You make it sound ominous."

"Oh, that was not my intention. I just mean that everything may not be as cut-and-dried as it's being made out to be. That may be silly of me, a mere prejudice against the obvious. It goes back many years to my training in . . . the devious." He leaned back in his chair. "But let us stick to the point—your two bothersome enigmas."

"Yes. Have you any explanations?"

"Oh, certainly. I should have thought by now the explanations would have suggested themselves to you. They are reasonably obvious."

"Then I'm being exceedingly dense."

"Perhaps. But I prefer to think of it as loyal."

"Loyal? Now I really am puzzled."

Dobbs leaned back even farther in his chair, lacing his hands behind his head. He spoke slowly, his words drifting up into the trees, as if he were addressing one of the birds that fluttered amidst the leaves. "Your reluctance," he said dreamily, "to reveal to Paul Ostrow that Harrington entertains an alternative view—that the killer may be neither the KGB nor the much-vaunted intruder. Of course, as you have suggested, a word to Paul is a word to Emory. So it is from Redwood that you are keeping Harrington's dark hints. And why, you ask. *Because*, the answer comes back loud and clear, you rightly suspect that Redwood would understand at once what it is that Harrington is hinting at. And you don't want to give Redwood that opportunity."

"But I don't even know what it is Harrington is hinting at."

"Sure you do. You're just repressing it."

"Okay, Dr. Freud, why am I doing that?"

"I told you. Loyalty."

"I don't get it."

Dobbs straightened up in his chair and glared at me. "Howard, you are wilfully behaving like a classic textbook case. If the murder was not committed by the KGB, and not by an inopportune intruder, then clearly Harrington is suggesting that it was committed by someone at Redwood Press, someone in your office."

He got up and walked toward the house. "Care to watch some television?" he asked. "Sunday is my evening for television. I love to watch Alistair Cooke and snarl at his damnably perfect delivery. It makes me feel humble, and I suspect that's good for me."

For a few minutes, however, I just sat out there alone in the gathering gloom.

10

Going to the office from the Port Authority was like coming into Istanbul, only more depressing. The walk across town from Eighth Avenue to Times Square and over to Grand Central reminded me of the seamier streets of the Turkish city, except that I'd felt safer in Istanbul. The crowds in the terminal streaming down into the subway entrance prompted me to forgo the shuttle and walk across to the East Side. This was a mistake on a sticky Monday morning in summer. My uneasiness was compounded by the self-consciousness I felt in carrying my bag. I felt like a tourist, which no New Yorker cares to be mistaken for. And I also felt like an easy mark for every derelict, drug addict, pimp, prostitute, and lowlife lurching out at me from doorways, storefronts, and parking lots. The only missing reminder of Istanbul was the beggars, and that seemed to be simply because the potential beggars were sleeping on the sidewalks at that early hour.

The stink of rotting food was mixed with gasoline fumes and a pervasive bathroom odor. Along with the smells came the noises—buses and trucks shifting gears, horns honking singly and in chorus, the squealing of brakes and tires, and, all across Forty-second Street, the insistent clamor of sirens from police cars and ambulances, even at eight forty-five in the morning. For me, this was an unaccustomed look at the city. The regular commuters seemed to ignore it. It was a relief to enter the subway at Grand Central and head downtown for the office.

Milt, Myra, and Mr. Snap were all at their desks when I arrived, and I nodded to each on the way to my office. None of them looked suspicious or murderous, but then I only glanced at them fleetingly from the hall. Dobbs's revelation of what Harrington had been hinting was a crusher. How was I ever going to face my co-workers, knowing that the CIA had marked one of them as a murderer? Of course, that was an exaggeration. As Dobbs had pointed out—repeatedly—this was only what Harrington was hinting at; he hadn't said it was so; he might not even truly believe it; he could be wrong. Also, he could be referring to someone in the bullpen or the mail room; he might be hinting at Emory, which could be just spite. It was all very tenuous, that was Dobbs's word, but it all came down to the same thing. The CIA didn't believe the KGB theory; that I knew. And it presumably didn't believe the intruder theory. The CIA preferred to believe that one of the people I worked with was Minnie's murderer.

Which one? Milt, Myra, Paul, or Sherwood? Or was it Emory? Or Mr. Snap? Good heavens! It was Roger McGraw. That smooth, cold, emotionless efficiency machine was a perfect suspect. He was the least likely person. That made him the most likely person. That's how the CIA mentality worked. My mind was racing along these Byzantine channels as I hung up my jacket, opened my window, and flipped through the mail on my desk.

An envelope from a well-known publishing house caught my eye. Above the printed return address in the upper right-hand corner, the letters DF had been typed in. I tore open the envelope and eagerly read the brief note it contained.

> I am thinking of preparing another meal for you. You were so appreciative of the first one, I can't resist. Besides, you look like you could use some good home cooking. Name the day.
>
> Your Cook

I grabbed one of my letterhead memo pads, tore off a sheet, and wrote:

Any day. Only this time, make it at my place. You'll need to get in beforehand to prepare it. So here's a key to my apartment. Oh, subtle, subtle.

I included my address and apartment number, and signed it, "Your Gourmet." In one of my frequent excesses of caution, I had made extra keys to my apartment, one of which I kept in my desk drawer in the office. This key I taped to the note and slipped into an envelope, which I addressed to Dinah at her office. The envelope I marked "Personal" and dumped into my "out" box. That taken care of, I sat down at my desk and waited for the usual Monday-morning parade of visitors.

Milt or Sherwood could usually be counted on for a quick chat, and Paul invariably office-hopped on Monday mornings. We all sort of checked in with each other, to catch up on the weekend's activities. It was a ritual. Except this Monday morning. I stayed put at my desk, and so did everyone else, apparently. I couldn't decide whether I was just as glad, or not. I worked away at catching up on correspondence, but I kept one eye on the doorway.

Late in the morning, Emory stopped by. "I'm on my way to Washington," he said. "You don't need me for anything, do you?"

He looked like he didn't really want an answer, so I just shook my head, and he was off. I heard him stop by Mr. Snap's office, but only for a minute. It looked like a rush visit, probably a summons from his masters to explain his annoying behavior.

Just before noon, the phone rang. I picked it up and said hello without paying much attention.

"Miller!" The peremptory tone and the fact that, even from one word, I recognized the voice as Harrington's snapped me alert.

"Yes, it's me."

"Has Redwood left yet?"

"Yes, about half an hour ago."

"All right then. Stick around this evening. Don't go home after work. How late does the office empty out?"

"Everyone's usually gone by five-thirty."

"Good. I'll be in about five thirty-one. Is the door from the reception area left open?"

"Sometimes. The last person out usually locks up. If the door is locked, there's a night bell. Just ring and I'll let you in."

"Five thirty-one," he said and hung up.

I was mulling over the surprise call—obviously, Harrington had arranged for Emory to be away, or at least knew he would be in Washington—when the phone rang again.

"Hello."

"Dear boy," came an even more familiar voice, "have you survived the return journey?"

"The trip was fine," I replied. "It's just that coming into New York at the Port Authority is a bit of a culture shock for a confirmed East Sider."

"Imagine how it is for us country folk. I don't inhale until I get past Fifth Avenue."

"I'm thinking of writing a piece comparing Eighth Avenue and West Forty-second Street to Istanbul."

"Which one comes out better? Or shouldn't I ask?"

"I haven't decided."

"Save it for your memoirs. I'm going to be in your neighborhood later. I thought I would stop up at your office. I don't want to bother you during working hours. Will five-thirty be all right?"

I hesitated, flustered. "I'm expecting someone at five-thirty."

"Good. You can introduce me. I'll put on my best company manners. Five-thirty, then." And he hung up.

What was going on? Dobbs wasn't usually so abrupt. And he hadn't even asked who it was I was expecting. I wondered if he knew. Had he and Harrington been in touch with each other? And if so, why? Oh, well, I thought, I'll find out soon enough. Curiouser and curiouser, as Alice had said about *her* Wonderland.

I had all I could do to manage lunch and a normal afternoon's work. Now I was sure I didn't want to chat with anyone in the office. But Emory's absence was invariably a signal for everyone at Redwood Press to wander around the premises.

Paul came by to inquire about my weekend and to say we must have lunch more often. I told him my weekend was quiet and that, sure, I'd love another lunch sometime soon. I had to see Roger about a production problem and ran into Sherwood on my way to his office. Sherwood said I was a stranger, and I replied that it was always much too warm on his side of the office. Even Roger wanted to chat, which was unlike him and drove me to distraction. Why is it that when you are particularly disinclined to be convivial, the least chatty person on earth decides to jabber away like a long-lost relative?

A visit from Milt was inevitable, and even Mr. Snap stopped by to find out whether my weekend had been pleasant. I answered them with generalities, determined by this time to keep my stay at Dobbs's place a secret. Ordinarily, I would be bubbling over with news like that, but now, for some reason, I felt self-conscious. Or was I sensitive about Dobbs and Harrington? I was puzzling over that question when I bumped into Myra in the hall.

"My, you look preoccupied."

"Just some work," I said. "It's all piling up. I think I'll stay for a while this evening and get it done."

"Oh, good," she said. "I was thinking of staying late one evening, but I hate to be here alone. If you're going to stay, then I'll do it tonight."

"Oh," I said, shaken. "Don't stay on my account. I mean . . . don't count on me. I'm just going to stay about an hour."

She looked at me curiously.

"I just didn't want you to make plans . . . for a whole evening, I mean. I'm really not staying that long. Maybe," I continued lamely, "we can make a date for later in the week."

"I'll consult my date book," she said, stalking into her office.

I ran back to my office in a sweat.

But everyone was indeed out of the office by five-thirty, except for me, sitting at my desk, trying to look like I was working. I had my radio on, and the five-thirty traffic report had just been announced when Harrington appeared in my doorway.

"The door was open," he said.

"Did you leave it open? I'm expecting someone."

"I know. I spoke to him earlier."

"Did you invite him?"

"It sort of came up when I said I'd be in to see you. He likes your company, says you have lots of interesting things to say."

"I didn't know that you knew each other," I said petulantly. "You seemed to be annoyed the other day, when I mentioned that I had been speaking with him."

"I got over that soon enough, didn't I?"

"Do you know him?"

"I know about him." He smiled cryptically. "Apparently more than you do."

"What's that supposed to mean?"

"Oh, the war . . . and all. No. I don't know him. But I'd like to meet him. We had a nice chat on the phone earlier today."

"Well, he'll be here," I said, thoroughly mystified.

Harrington sat down, adjusting himself to the lumps in the chair and muttering.

"It's meant to discourage visitors," I remarked.

"Is Redwood's the only decent office in this place?"

Before I could answer, Dobbs appeared.

"I locked the door behind me," he said. "I thought we would prefer to be alone. We are alone, I take it?"

Harrington stood up and extended his hand.

"Myles Harrington," I said, "Hartley Dobbs."

"A pleasure," Harrington acknowledged. "Your thing about Beirut is still required reading around our shop."

"You flatter me," Dobbs replied, shaking hands. "Nowadays, I confine myself to things more . . . entertaining."

"I thought you were going to say 'more lucrative.' "

"That too." Dobbs looked around and said to me, "We need another chair."

Harrington motioned him to the armchair, but he looked at it scornfully and went out into the hall. In a moment, Dobbs was back. "I'll just wander around, or sit on the windowsill."

Harrington sighed and resumed his seat. "The question I want to clear up first," he said, "is who might have known what about the two manuscripts."

"That's no problem," I said. "Emory, Fran Bishop, Paul Ostrow, and I knew about the two manuscripts. Emory because he concocted the whole idea; Fran because she's his secretary and takes care of all sorts of things for him; Paul because he was going to go to Istanbul before I got into the act; and then, of course, there's me. That covers it."

"Well, not exactly," Harrington replied. "There's me, and a few others I could name. Of course, we just sat around while Redwood concocted the whole idea. But we knew about the two manuscripts. And there's our friend over here." He waved an arm at Dobbs. "He knew about the two manuscripts because a little birdie told him. The thing is, were there any other little birdies talking out of turn?"

"To whom?" Dobbs asked him. "Do you have any prime suspects?"

"What about Snap? Would Redwood have told him? They are partners, you know, at least in a business sense."

"Maybe in a business sense," I argued, "but not about editorial matters. It's very rare for Emory to include Fred Snap in an editorial decision."

"This wasn't exactly an editorial matter," Harrington noted.

"More so than a business matter," I persisted.

"What do you think, Dobbs?"

"I think you might sit down and talk to Fred Snap. From what Howard has told me, I sense a man much frustrated by the conditions of his business partnership. I would also hazard a guess that he knows more about what is going on—in general—than his employees seem to think."

"You think I should talk to him."

Dobbs perched himself on the edge of my desk, facing my shelves. He was sideways to me and had his back to Harrington, fidgeting in the armchair. "Ask him whether he knew about the two manuscripts," he said over his shoulder. "It can't hurt. And you may be surprised."

"Nothing would surprise me," Harrington replied. "Miller, is there some way one of the others could have known in advance about why you were going to Istanbul?"

"Milt . . . Sherwood . . . Myra . . . Roger . . ." I ticked them off on my fingers. "No. I don't see how they could have known. And I didn't tell them, if that's what you're getting at."

"You left someone out," Harrington said.

I started counting on my fingers again.

"He means Minnie," Dobbs said, without turning around. "That's an interesting idea, Harrington."

"Well, it opens up all sorts of possibilities. There's a loose end here that I'm trying to tie up. Only I can't seem to get my hands on it."

Dobbs got up off the edge of my desk and walked over to the shelves. He began riffling the pages of the many manuscripts and galleys piled there. "How can you work in this mess?" he asked me. "How do you know where anything is?"

"It's not a mess," I countered. "I know what's there. Besides, it isn't really important anymore. It's only dead matter."

"Dead matter?" He looked at me. "What's that?"

"They institutionalize their chaotic practices in this business," Harrington told him. "Every little mess that they make has a special name. If they can't keep track of it, they call it 'slush,' or something like that."

"Slush I understand," Dobbs said. "What's dead matter?"

"When the printer is through with a book," I explained, "he sends back the manuscript and the various stages of galleys and page proofs. It all comes back in a package marked 'dead matter.' "

"That seems like drastic terminology."

"Some printers call it 'foul matter.' "

"Well . . ." Dobbs considered this. "At least it's less final than 'dead matter.' "

"We're supposed to return the original manuscript to the author," I said. "I usually do, eventually, around the time the book is published. By then we can throw out the galleys."

"You learn something new around here every day," Harrington said. "Are you as haphazard with the dead matter," he asked me, "as you are with the slush pile?"

"Look," I said defensively, jumping up and going over to the shelves, "this is not chaotic. I know what every piece of 'chaos' is on these shelves. It may look like a mess to you. And maybe I should have tossed out some of this stuff ages ago, but I know . . ." My voice trailed off.

"What's the problem?" Dobbs asked.

I stared at the shelves questioningly. "I don't quite . . . there's something . . . wait a minute."

"Yes?" Dobbs queried.

"The *gestalt* isn't quite right."

"Oh, the *gestalt*," Harrington said. "Is that another publishing term for 'confused mess'?"

I ignored him and went over to a shelf on the right. "I don't quite remember this," I said, reaching for an envelope lying among three others. It was a large brownish envelope, manuscript size, and, indeed, it was open at the short end, revealing the pages of a manuscript. Both Harrington and Dobbs watched in silence as I removed it from the shelf and extracted the pile of typewritten pages. "This paper . . ." I mumbled, "seems wrong. . . . I never had a manuscript on this kind of . . . Oh, my God!" I gasped, staring in disbelief at the manuscript in my hands.

"What is it?" they asked in chorus.

"Is this what I think it is?" I shoved it at Harrington. But I knew. It was in French and an original copy. I had only glanced at the title page and a few of the inside pages. There was no author's name on the title page, and my French was rudimentary. But I knew.

Harrington took it as if it were a holy relic. He examined the top page and flipped through the rest, pausing occasionally to read a few lines, then handed the manuscript to Dobbs. "Shit!" he said, with extreme feeling.

"I take it," said Dobbs, "this is the French Kuzatov manuscript."

"You take it right," Harrington stated. "Shit, shit, shit!"

"Your vocabulary seems to have come unhinged. Can you tell if it's genuine? You only glanced at it."

"I'll take a close look later, but it looks right. The thing is, what's it doing on Miller's shelf? And how long has it been there?"

Dobbs handed him the manuscript. "Why so perturbed? I should think you'd be overjoyed to find it there."

"Overjoyed?" I snapped back. "Why should he be overjoyed that I found the goddam manuscript on my shelves?"

"Calm yourself, dear boy. You're obviously not following the line of reasoning." He pushed me back down in my chair and perched himself on my desk. "Now, what did I tell you yesterday about his hints?"

"You said he was hinting that someone at Redwood Press, someone in this office, might be the murderer." I turned to Harrington. "Is that what you believe?"

"Not necessarily Heffernan's murderer," he replied evenly. "Involved in the disappearance of the manuscript, and maybe in the murder, is more like it."

"You really believe that?"

"The evidence seems to support me," he said, holding up the manuscript.

"But you believed that . . . before."

"Howard," Dobbs said, holding up a cautionary finger, "you've not been overbright about this from the start. Perhaps that's a concomitant of your publishing experience. But consider things from Mr. Harrington's point of view. A manuscript doesn't show up where it's supposed to; then it shows up where it's not supposed to; only nobody sees it until it shows up—and promptly disappears—at the scene of a murder. Because this kind of thing is readily explainable to you in terms of transoms and slush piles and other peculiar publishing practices, you don't see something that must be all too obvious to Mr. Harrington." He glanced over at the CIA man, as if to see that he was getting it right. "All this activity that seems to you like a series of understandable, explainable accidents, smacks of a guiding hand to him."

Harrington nodded his head and grunted.

"And now," Dobbs continued, "the manuscript shows up at last, in the offices where the guiding hand is most likely to be found."

"What guiding hand?" I looked from one to the other of them.

" 'Guiding hand' is his phrase, not mine," Harrington said. "I thought of it in terms of *'convenient'*—once that manuscript didn't show up in Paris, how convenient that it was safely tucked away in these offices; how convenient that it was gone as soon as we knew where it had been."

"So, if you follow that," Dobbs put in, "you can see that the convenience centers itself upon these premises. And now the manuscript shows up here."

"But I didn't put it there," I practically shouted.

"Of course not, that's the beauty of it. The manuscript is always *not* where it should be . . . or, to put it the other way, it *is* where it shouldn't be."

"I don't think I follow that."

"But I do," Dobbs said, as if that were all that mattered. "And that's why I ask our friend what's perturbing him."

"Doesn't it strike you as too neat?" Harrington asked.

"It's consistent," Dobbs argued.

"It was, up to this point."

Dobbs considered this. "Yes, I see what you mean. It's confusing to have it show up again, under these particular circumstances. What do you know of the police investigation? Certainly, from what I gather of their suspicions, this latest discovery won't please them."

"I haven't had my sit-down with Preston yet. Of course, now"—he waved the manuscript at us—"a long talk is mandatory. And with this to offer, I won't have any trouble getting in to see him."

"But you have some idea of how the investigation is going."

"He hasn't really given me anything. But I know Preston slightly, enough to say that if he's sticking to his intruder theory, it's because he has something good enough to substantiate it—or something that looks good enough for the time being. Preston would just clam up, rather than go out on a limb."

"Some people here," I volunteered, "think the police are fixing on an intruder as a convenient scapegoat."

Harrington mulled that over and smiled. "I smell the Viennese school of philosophy there."

"No," I said, "not just Emory. Most of the people in this office think the police are baffled and are just using the intruder as a convenient out."

"That may be," Harrington replied. "But my guess is, if Preston was really baffled, he'd just keep his mouth shut."

"Then the presumed police line," Dobbs said, "doesn't jibe with yours. Though, of course, you indicated that already."

"I don't have a line," Harrington countered.

"Don't quibble. You already admitted your uneasiness over what you call 'convenient' circumstances. But I'll offer you another alternative to consider. What if the manuscript and the murder are not strictly related?"

"That's silly; of course they're related."

"I mean in the strictest sense," Dobbs went on, unruffled. "What if retrieving the manuscript was one thing, involving one person or set of persons, and murdering Minnie was separate, involving another person—an intruder, say, or the KGB?" He shook his head. "No, that's not right. The KGB would be after the manuscript. What if . . . ?"

"Are you buying that KGB nonsense?" Harrington interrupted.

"I'm trying to think why the manuscript got back here."

Harrington stared at Dobbs glumly, then began pacing in the small space near the chair. He was thinking hard and absentmindedly riffling the pages of the manuscript. "No," he said at last, "even leaving out the KGB, it's too coincidental. I don't like that much coincidence."

"Coincidences do occur," Dobbs reminded him. "I think you may be too suspicious. After all, the question is why the manuscript got back here on Howard's shelf . . . and how. But I prefer to think that if we figure out why, we'll discover how."

"Too much for me," Harrington muttered.

"Yours not to reason why?" Dobbs asked with a grin.

"I'll leave that to you."

I was growing impatient with them. "Listen," I said, "that manuscript was on my shelves. I want to turn it over to the police. If you two weren't here arguing theories, I'd be on my way to the police with it now."

"Oh, of course it's going to the police," Harrington said. "I'm bringing it down to Preston myself, first thing in the morning. And one more thing, Miller . . ." He wiggled a finger at me. "Not a word of this to anyone, not to *anyone*—most of all, not to Redwood." He looked at Dobbs and nodded, as if for confirmation.

Dobbs nodded back at him and reached for the manuscript. "May I just examine it again for a minute?" he asked. Harrington handed it over, and Dobbs studied it for a few minutes, examining the title page and then reading at random.

"You'd better remind Preston that the three of us have gotten our fingerprints all over it." He turned to me. "Where's the envelope it came in?"

I handed it to him, and he took it gingerly. It was a plain brown envelope, of that odd size and kind of paper I automatically associated with Europe. Our envelopes were invariably lighter in color or gray, and of heavier stock. Dobbs turned it over. It had obviously once been sealed; the paper on the back had that rubbed-off look where strips of tape had been removed. The clasp was the kind where two little metal prongs are pushed through a hole and flattened down in opposite directions. Though the envelope was now open, the flattened metal prongs and rubbed paper clearly suggested that the envelope had originally been clasped shut and sealed with tape. We all examined this, touching as little as possible. Dobbs flipped the envelope again, and we examined the address side.

No return address was indicated, but there were French postage stamps, thoroughly and indecipherably canceled. The address was handwritten—just Redwood Press, the street and city, U.S.A., and the zip code—and in the lower left corner, in the same hand, *"Par Avion."* Harrington peered closely at the postmark and mumbled something about "Paris, probably." He shoved the pages of manuscript back into the envelope, monkeyed with the prongs a moment, and then pressed it closed.

"An envelope marked like that," he asked me, "would have gone into the slush pile?"

"Most likely. It wasn't addressed to anyone in particular."

"Even though it had French stamps on it?"

"The mail boy might not have noticed, or cared. If one of us had seen it—for instance, if I had seen it in the slush pile—we might have picked it up. It has that different look . . . European . . . it's the paper. But don't count on that."

"I don't count on anything in this office. For instance, how long could this thing have been on your shelf?"

"As long as I've been back, I guess. Since the day it was removed from Minnie's apartment."

"You're just assuming that," he said disgustedly. "The way you people operate, it could have been there a year."

"Now that's not—"

"All right. The likelihood is that it was put there after it was removed from Heffernan's apartment. But you don't know that for sure. And if you hadn't discovered it just now, it could have stayed there another eighty years. And that's the fact of it."

"That's not fair."

"I wish you two would stop nattering and let me think," Dobbs said. "There's something about the envelope I don't like, but I can't put my finger on it."

"They specialize in that in this office," Harrington responded. "I've never seen anything so haphazard."

"Oh, well," Dobbs said. "It will bother me, no doubt, until it comes to me at two in the morning."

"My problem," Harrington said, "is trying to find out what went wrong in Paris. I've been following some leads, but investigation of them on that end is . . . well, embarrassing to the CIA. And one thing we don't need at the moment is more embarrassment. Dobbs, would you do us an enormous favor and get in touch with your friend Colonel Hapworth to conduct a discreet inquiry on our behalf?"

"The colonel is not exactly a friend, Harrington."

"Oh, you know what I mean. Use your connections. It would be enormously helpful in this . . . these circumstances."

"And I thought you wanted to meet me for the pleasure of my company."

"I'm sure your company is a pleasure," Harrington conceded. "I've seen you on television . . . and even enjoyed it. But this business is . . . well, let's say it's a pressing matter. I know some very specific questions I would like asked in Paris. But it would really be an embarrassment, as I say, and we would prefer some outside help. . . ."

"And you think Colonel Hapworth could provide it?"

"I'll tell you the questions and the people of whom we want them asked, and I think you'll agree."

I had been following this conversation like a man at a Ping-Pong game, turning my gaze from one to the other of them. Now it was getting too much for me. "What is all this Alphonse and Gaston routine?" I asked petulantly. "And who is Colonel Hapworth?"

"You'll have to excuse us a minute, Miller. I want to speak with Dobbs in private. You stay where you are." He waved me back into my seat. "We'll just walk down the hall."

"Hey!" I shouted. "What's going on?"

Harrington was ushering Dobbs into the hallway. He turned to me and said, "I asked you once before if you knew who Dobbs was—his background. Obviously you didn't know then and you don't know now. So he didn't tell you. I don't know if it's for me to say." He looked at Dobbs, who shrugged his shoulders and marched down the hall toward Emory's office.

"To put it very briefly, your friend was in British Intelligence during World War II. And I'm counting on the unwritten law that says nobody ever really leaves that service. So, if you'll just excuse us, we'll be back in a minute or two." He followed Dobbs down the hall.

I sat back in my chair, blinking at the doorway. Dobbs had once been a secret agent . . . those unspecified World War II adventures, presumably. I couldn't conceive of it. And he had never even hinted at it. And his "thing about Beirut" that Harrington had mentioned—some classic account of spying or espionage, no doubt. I think I was still blinking at the doorway when they reappeared in it.

"I much appreciate this . . ." Harrington was saying. "*We* much appreciate it. Now, what was it you wanted to suggest about this business?"

"I have a little idea," Dobbs replied, "for smoking out whoever it was who removed the manuscript from Minnie Heffernan's apartment and placed it on Howard's shelves . . . the person who may or may not be Minnie's murderer."

"What's that?"

"Where is the garbage dumped?" Dobbs asked me. "I mean the trash, the paper rubbish, contents of wastebaskets, and things like that."

"There are some rolling dump carts that are kept in the back, in the shipping area. The cleaning people come in around eight o'clock and dump everything into those carts; then they wheel them out to the freight elevator and take them down to the street. A private carter comes early in the morning and dumps the stuff from the carts into a garbage truck. During the morning, our mail room people bring the dump carts back upstairs."

Dobbs seemed impressed.

"You seem to know a lot about it."

"If you work late enough times, you see them do it and catch on to the routine."

"That's the most organized thing I've heard about this place," Harrington remarked. I ignored him.

"Then the dump carts should be out in the shipping area."

"Yes."

"Show me where that is. Come, Harrington, we'll probably need three of them."

Harrington followed us out. "If you have in mind what I think you have in mind," he said, "I don't think it'll work."

"Why not?" Dobbs argued. "At the very least, it's going to make someone extremely jittery. And a jittery thief, or murderer, will play right into our hands, or rather Lieutenant Preston's hands. You'll tell him what we've done."

"What *are* we doing?" I asked.

"We're going to empty your shelves, Howard. We're going to clear them of every last blessed scrap of paper."

"You can't do that!" I cried.

"Why not? It probably should have been done ages ago. You, yourself, as much as admitted that. Now it won't be so painful. Cleansing is good for the soul. And if there's anything that really needs saving, we'll find some place for it. Ah, here we are."

We were in the shipping area, and he stopped in front of the four rolling dump carts that were waiting for the cleaning people.

"I think two will be enough," Harrington said.

"I agree," Dobbs replied. "Stop jumping up and down, Howard. You look like you're performing a Mexican hat dance."

"But you can't empty my shelves."

"Why not?" he asked again, with great deliberation.

I looked at him, trying to find an answer.

"See . . . you have no reason to give, except an unremitting resistance to change. Come . . . push one of these things back to your office, and try to look upon this as the first step toward a more orderly life."

"But what will this accomplish?" I asked, reluctantly pushing one of the carts back to my office.

Dobbs wheeled another cart behind me, helped by Harrington. "What it will accomplish," he said, "only time will tell. What it is intended to accomplish is to agitate whoever placed the manuscript on your shelf."

"Ah . . . agitate," Harrington mumbled.

"Well, it was a remark of yours that gave me the idea "

"A remark of mine?" Harrington was puzzled.

"Yes. You said the manuscript would probably have remained on that shelf for eighty years if Howard hadn't been goaded into examining the chaotic mess he keeps there."

"And so it would have."

"No doubt," Dobbs concurred. "Which is probably why it was put there in the first place. I suppose you realize—if you're following my train of thought—that this is one more argument in favor of your . . . suspicions."

"Oh, I'm following," Harrington said with satisfaction.

"I'm not," I said.

"You're supposed to lead. Just get us back to your office."

We left the carts in the hallway outside my office door, and Dobbs began pulling at galleys and manuscripts on my shelves. "I'll hand these things to you," he said. "If something absolutely has to be saved, or is an original manuscript that should have been returned to its author, poor devil, then you can put it aside. Otherwise, you will hand it over to Mr. Harrington, who will deposit it in the dump cart. We are going to empty your shelves, dear boy. The shelf's the thing wherein we'll catch the conscience of the . . . well, we're not really after a king."

He seemed sublimely pleased with himself, and there was absolutely no stopping him. Harrington seemed content to go along with his plan—or was indifferent, I couldn't tell. Helplessly, I passed manuscripts and galleys from Dobbs to Harrington, occasionally rescuing one and putting it on my desk for safekeeping. In less than half an hour we were done; the shelves were empty and the carts were nearly full.

Dobbs looked at the few manuscripts I had rescued. "Pile those neatly on the floor next to your desk, or over there on that table," he said, "and in the morning return them to their rightful owners. Meanwhile, we will return these chariots to their accustomed places." He stepped into the hall and began pushing a cart. "Come," he said to Harrington, "let us sing as we go. Do you know the 'Pilgrims' Chorus' from *Tannhauser*?" His voice floated across the empty office space as I piled the manuscripts on the table.

"What do we do now?" I asked when they returned, dusting their hands.

"Now," said Dobbs, "we go out to dinner; someplace where we can continue our discussions without too much interruption."

"I don't mind the dinner," Harrington said. "In fact, it's on me. But I would rather curtail the conversation until after I've had a chance to go over things with Preston."

"Fine," Dobbs agreed. "Then we can go to an excellent restaurant where our discussion shall be limited to purely innocuous matters, and where friends of Howard are treated like royalty. Call them up, dear boy, and see if they have a table at a moment's notice."

"What about my office?"

"A dramatic improvement."

"Never mind the jokes. What am I supposed to do, now that my shelves are all empty?"

"We'll explain to you over dinner how you're to act tomorrow morning. It should be a piece of cake for a trained operative like you."

I groaned.

"Buck up, Miller," Harrington told me. "Just make believe you're in one of those movies you're so fixed on."

"I'm glad everyone's in such a good mood at my expense. What about Kuzatov's manuscript?"

"I have that," he said, patting it. "And I'm keeping it with me till I can turn it over to Preston tomorrow morning. Come on, Miller, phone the restaurant and make a reservation. Then let's go wash. Your friend here has literally involved us in dirty work." He smacked his hands together. "You know," he said to Dobbs, "this is the first time I've had an appetite since last Monday."

11

My day in the office on Tuesday was not a piece of cake, as Dobbs had forecast, nor was it like any movie I had ever seen. At dinner the previous evening, Dobbs and Harrington had been breezy to the point of exasperation. Harrington was buoyed by having the manuscript to show to Preston; Dobbs was buoyed by the prospect of ferreting out the person who had removed the manuscript from Minnie's apartment. The buoyancy of one infected the other, and between them they were positively ebullient. And the higher their spirits rose, the more depressed I became. They talked about cities they had been in and enjoyed, asking me how their descriptions stacked up against Istanbul. They ordered wine and more wine, and I, a poor wine drinker, developed a headache and a knot in my duodenum. Both were still with me when I arrived in the office on Tuesday morning.

"Just act like nothing happened," Harrington had said. "You can do that."

"You stayed late and cleaned out your office," Dobbs had said. "You just got tired of the mess and chucked it all out."

Their breeziness was not moving me, except to annoyance. "You didn't look at anything," Dobbs said. "You just chucked it all into the garbage."

"There was nothing there worth looking at," Harrington said. "There was nothing worth keeping."

And that was the extent of their advice. I was to sit in my office and note everyone's reactions to my empty shelves. When I raised objections about possible questions, they both shrugged them off. I would be able to handle it. Nobody would ask embarrassing questions. I worried too much. Piece of cake. Play it like Humphrey Bogart.

A little before nine-thirty, Humphrey "Piece of Cake" Bogart received his first visitor.

"Good morning, merry sunshine," Milt boomed.

I cast him a bilious glare.

"What's the matter with you?"

"I'm not feeling well. My stomach bothers me. I have a headache."

"Probably the result of bad living," Milt said cheerily, seating himself in the lumpy armchair. He glanced past me toward the shelves.

"You've done something to your office."

"What makes you say that?"

"I know. You've emptied your shelves."

"How clever of you to notice. Tomorrow I'm getting rid of that chair."

"Good idea," he said. "I mean the shelves. I keep promising myself I'm going to do the same thing. What'd you do? Just chuck it all into one of those bins?"

"That's right. Lock, stock, and barrel."

"Very smart," he said, getting up. "Well, I can see you're not going to be good company. You ought to try an Alka-Seltzer. If you want one, I have some in my office."

That was that, and pretty much a pattern for the rest of the day. I sat at my desk, making an attempt to work, but really intent on registering everyone's reaction as he or she came in to chat. They were all, however, replays of the scene with Milt. Mr. Snap commented that I had introduced a new hazard for editors—hernia. Sherwood assured me that I had undoubtedly thrown out something I would need the next day. Emory took much the same tack, but decided it was a good idea to clear the shelves and would recommend it to the others; he himself never kept anything unnecessary. Roger showed up to discuss some photographs for one of my books, glanced at the shelves, nodded his head appreciatively, and stuck to business. Only Paul and Myra came close to exciting my antennae.

"What have you done to your shelves?" Paul asked.

"I've emptied them. I should think that's obvious."

"When did you do that?"

"Last night. I stayed for about an hour or so and just dumped everything into those carts we keep for the garbage."

"Probably something we could all do," Paul agreed. "What did Emory say?" Considering the juxtaposition of those two sentences, I decided that Emory had sent him into my office.

"Emory said it was a good idea," I replied. "He'll probably order a general housecleaning now. I'm sorry if my sudden zeal inconveniences everybody."

"No matter. What was on the shelves?"

"Just a lot of dead matter. There were one or two original manuscripts I knew about. Those I returned to the authors this morning. But the rest I just removed by the armful and dumped in the carts."

"Last night?"

"Yes."

"Then it probably all got taken away early this morning. I hope there was nothing important that you overlooked."

"I don't think so."

"Well, let's hope not," he said, and walked away.

Myra passed by my office in the afternoon, looked in, and spotted the empty shelves.

"Then you did stay last night," she said.

"Just for a little while."

"You seem to have made good use of your time." She waved an arm at the shelves.

"I just thought it was something that needed doing."

"In your case, long overdue."

"I suppose so."

"Did you have to go through everything first? That must have taken some time."

"Oh, no. There were one or two things . . . manuscripts that had to be returned to authors. The rest I just chucked."

She nodded her head thoughtfully. "Wholesale housecleaning. Good idea. No second thoughts. No regrets."

"What's to regret?"

"Oh, you know . . . the squirrel mentality. We all seem to have it around here. You start mulling over something, a manuscript or a galley, and the next thing you know, you've kept it all over again. You're back to the same mess you always had."

"No," I said. "I resisted that. Just took things out by the armful and dumped them."

"Where?"

"In the dump carts."

"Then they're probably all gone by now. They empty those carts in the morning, first thing." She surveyed the empty shelves. "Good idea . . . all gone . . . no regrets."

And that was it. Nobody had looked at the empty shelves, gasped, staggered, and fallen in a faint. Nobody had asked if I'd happened to run across a funny-looking French manuscript in a brown envelope. In fact, nobody seemed much interested, as far as I could tell. So much for Dobbs and his hotshot ideas.

I wanted to tell him about how it had gone, while I could still remember the conversations. But he wasn't available; he'd probably gone to Washington, or wherever, to see Colonel What's-his-name. And I didn't know how to get hold of Harrington. It was very frustrating. At five o'clock, I settled for calling Dinah.

"Oh, sorry, sweetie," she said. "I can't stop to talk now. I'm off to a screening. Just as well, anyhow. We're probably seeing too much of each other. 'Bye." And on that unsatisfactory note, she hung up.

On top of all these frustrations, it had started to rain. There wouldn't even be a ballgame to watch later. I went home wrapped in a gloomy and a comfortless mantle of self-pity.

On Wednesday morning, while showering and shaving, I contemplated the day's activities and realized that it was editorial meeting day. That realization was almost enough to send me back to bed. The last thing I wanted was to be locked up with my co-workers for an hour or two. As it turned out, the experience was even more frustrating and depressing in actuality than it was in anticipation.

Emory was in his most ebullient mood, and Emory bubbling over is about as enjoyable as Vesuvius bubbling over. He strode into the conference room exuding a self-satisfaction entirely at odds with the seven depressed psyches around the table. For one thing, the empty seat to my right was a forceful reminder of Minnie. For another, none of us had any particular project to bring up. The absence of projects was no dampener of Emory's spirits. He never even gave us a chance to display our lack of activity.

"Well," he said, barging through the doorway, "all my chickadees are here." He beamed at us, seated himself, and rubbed his hands together delightedly. "I can announce now," he said, "a paperback sale of the Kuzatov for a comfortable six figures. There is a press release going out right now, and copies of it will be on your desks before noon. I am sure you will all be as delighted as I am with the sale."

"Did you get what you were hoping for?" Paul asked.

"Just about. The escalators will more than make up for the difference. So it behooves us all"—he glanced meaningfully at Milt—"to push the sales of the hardbound edition as vigorously as we can. The production schedule"—here he shifted his gaze to Roger—"has been tightened to allow for an early publication date. But I'm sure everything will move smoothly on that end. Mr. Leitner"—he turned to Sherwood—"you and I had better meet this afternoon to go over the promotion. Yes," he continued with a smirk, "it was an excellent negotiation. Even the great Dinah Foxworth couldn't have done better."

Mention of Dinah's name was all I needed to cap my depression. Why had she been so flighty yesterday on the phone? What the hell did she mean about seeing too much of each other? I wished that the meeting would end, wished that it hadn't even taken place. I wanted to call Dinah, not sit in this stuffy room listening to Emory congratulate himself.

Emory beamed at us one by one. "Who would like to begin?" he asked, the Viennese dentist surveying his reluctant patients.

Sherwood raised a tentative hand.

Emory nodded and waved at him to proceed.

"I have a suggestion," Sherwood mumbled in his most apologetic tone. "I think it would be good public relations."

"Go ahead."

"I think maybe we should go back to the agent . . . what's her name . . . sort of as a tribute to Minnie . . . go back to the agent and make an offer on the Audrey Burbage book . . . you know, the one that Minnie was so high on."

His voice dribbled off and a heavy silence hung over the room. I glanced warily at Paul, who was staring down at the table, and at Myra, who was gazing stonily at some space a few feet in front of her, Minnie's empty seat perhaps. Finally, I turned to Emory. A deep color had suffused his face; his eyes had narrowed and his lips were pursed. Vesuvius had stopped bubbling and was about to erupt.

"What do you think, Emory?" Sherwood asked hesitantly.

"I think," he replied in measured, expressionless tones, "that that is one of the stupidest suggestions I have ever heard."

"Why?" Poor Sherwood. He gave up on his good ideas in the face of any opposition, yet invariably chose to defend the ones that had no hope of winning favor.

"You ask me why?" Emory responded between clenched teeth.

"It would be very good public relations . . . and good publicity. We could publish it as a 'Minnie Heffernan Memorial Volume.' "

Emory's reply was a sound like a motor rumbling.

"Sherwood," Paul put in quickly, hoping to defuse the situation, "I recognize your good intentions . . . but your suggestion seems to be a bit tacky."

"Tacky?"

"Yes. Perhaps you misunderstand me. Let's say commercial in a bad-taste way." Paul had been puffing on his pipe, and now he coughed. "Yes . . . that's what I mean. It would be looked upon as our trying to commercialize poor Minnie's death."

"We owe her *something*!" That was from Myra. She was still staring at a space in front of her, and her voice had an oddly sepulchral sound. I was reminded of Lady Macbeth's sleepwalking scene.

"However that may be," Emory said, his anger just barely in check, "we do not owe to her memory a book that we would not publish when she was alive."

"*You* would not publish," Myra countered.

"What is that supposed to mean?"

"Minnie would have published it. She wanted very much to have it published . . . and to be associated with it."

"Not under this imprint," Emory snapped. "If that is the tenor of the conversation this morning, I think we would do better to forgo the editorial meeting." He glared at each of us in turn. "Unless someone has a sensible proposal for discussion."

Everyone was busy examining the grain of the table. Emory snorted, flung himself out of his chair, and stalked out of the room. The rest of us sat listlessly for a minute, until Mr. Snap, murmuring, "Oh, well . . ." got up and wandered out. Roger and Milt quickly followed him.

"It was a kind suggestion," Paul said, gathering up his pipe paraphernalia, "but not a very sensible one . . . I mean . . . considering the circumstances." He walked out, shaking his head and sighing.

I looked across at Sherwood, who smiled and shrugged his shoulders. "Thanks, Myra," he said, "for sticking up for me."

She was still doing her Lady Macbeth number. "Not for you," she said. "For Minnie. I've felt so strongly . . . ever since it happened . . . that I owe her something, something to get her a little of her own back."

Sherwood and I left her sitting there, staring sadly into space.

As soon as I got back to my office, I phoned Dinah. She was in a meeting, her secretary said. Did I want to leave a message? I left my name. Did she have my number? Oh, boy, did she have my number! I said I was sure she had my number, and hung up. Then I tried Dobbs at his home. There was no answer. Frustration after frustration.

The first break in my gloom came in the afternoon and from an unexpected source. I was sitting at my desk, reading a manuscript, a measure of my desperation, when the phone rang. It was Harrington.

"Things are looking up," he said cheerfully.

"I'm glad to hear it."

"You sound down in the dumps, Miller. Cheer up. I told you, things are looking up."

"Wonderful. I'm happy for you. You're running around like Pollyanna, and I'm thinking of renting myself out to a Russian novel."

"What's eating you?"

"Where's Dobbs? I want to speak with him."

"I've just heard from him. He's been marvelously helpful. I can't tell you how much. It was a lucky break for me that you know him. I really want to thank you."

"You're entirely welcome. Where is he?"

"He's on his way back from Washington. He told me to tell you he'd meet you at the restaurant—I guess he means the one we all ate in the other night—about six-thirty."

"I'll be there. Now . . . don't you want to know what happened?"

"What happened where?"

"Where? Here! Here in my office . . . with the empty shelves."

"Oh, that," he said offhandedly. "What happened?"

"Nothing happened. Doesn't that interest you?"

He considered this briefly and said, "Take it up with Dobbs. It was his idea anyway."

"What about the manuscript? What did Preston say?"

"He's still working on it. I've discussed it with Dobbs. He'll probably tell you what we know when he sees you."

"Aren't you joining us?"

"I'd love to, Miller, I really would. That's a first-class restaurant. But I have some things that I really have to do. Sorry to pass it up."

"Well, I'll miss you."

"Yeah, sure."

"No, I mean it. You've suddenly made me feel happy for the first time in days. I really feel I owe you at least a dinner."

"I'll take a rain check. God, you have ups and downs. What happened to the Russian novel?"

"I'm thinking of switching to a French farce."

He laughed. "There's no keeping up with the editorial temperament. Try to stay out of the doldrums between now and six-thirty."

After he hung up, I tried Dinah again. Miss Foxworth was in another meeting, but she had my message. I went back to the manuscript I had been reading when Harrington called.

The restaurant was empty when I arrived. Franco ushered me to a table near the kitchen, disappeared up front at the bar, and reappeared in a minute with my brandy and soda.

"Franco," I chided him, "I couldn't order a different drink here if I tried. I'm hardly in my seat before you've descended on me with a brandy and soda."

"What different drink would you order, Mr. Miller?"

I gave it some thought. "A Campari and soda?" I suggested.

"That's a good summer drink," he conceded. "Next time."

I sipped at my drink, and in a few minutes Dobbs arrived. He nodded to Franco, who nodded back, and settled down at the table with a long sigh. It was hardly out of him when Franco arrived with his martini. He smiled at Franco, at the drink, and at me.

"I was just telling Franco," I said, "that we're going to change our drinks."

"Good heavens. We are? And to what, pray tell?"

"Campari and soda."

"Dear boy, if you want to change your drink . . . well, by all means. But leave me out of it. I'm perfectly happy as I am." He took a delicate swallow and sighed again.

"Hard day?" I asked.

"Long day. Otherwise . . . piece of cake."

"I've come to suspect you when you say that. But maybe for you things are a piece of cake. Anyhow, you must have done well, because Harrington is absolutely euphoric."

Dobbs concentrated on his drink, but a twinkle left his eye. "Is he now? That's nice."

"Are you going to tell me about it?"

"Only a very little." He continued to pay all his attention to his martini glass. "We hanky-panky fellas tend to keep our bag of tricks close to the vest. Or am I mixing my metaphors?"

"You're being exasperating, is what you're doing."

"Ah, of course." He smiled apologetically. "Why don't we order another round of drinks? I'll put my thoughts in order and see what's for publication. And you tell me what happened yesterday when everyone discovered your denuded shelves."

"Nothing happened." I waved a hand at Franco and indicated that we wanted more drinks. "Nothing at all . . . a big zero."

Dobbs "hmmmed" at this. "Tell me about it . . . in detail."

I went over each visit briefly, but in as much detail as I could recall. I was just finishing up on Paul and Myra when Franco arrived with our drinks—a martini for Dobbs and a watery but vividly red concoction for me.

"What's this?" I asked.

"Campari and soda," Franco replied blandly.

"It looks like cough medicine."

"And tastes like it too," Dobbs remarked.

Franco stood by while I took a sip. I made a slight face but said, "It'll do. Is it popular in Italy?"

"More popular than brandy and soda," Franco replied and walked away.

We sipped our drinks and finally I said, "Well?"

"Dear boy, you are insistent. Well, indeed. For your edification—and I will be purposely vague, if you don't mind—what you want to know is that no word of the French manuscript was leaked here, and no word of the Russian manuscript was leaked in Paris. In short, the plan proceeded much as it was supposed to. My sources were unaware of any mixup in Paris, or that the French manuscript wound up here in New York. They simply assumed that the whole business of a manuscript arriving in Paris was a cover for your admirable escapade in Istanbul. On the other hand, the manuscript was supposed to arrive at I. Pierre by mail. Can you imagine that? I wonder what the French equivalent of the slush pile is called. Harrington assures me that was the drill . . . by mail. Oh, well. The less I burden you with these matters, the better for your peace of mind."

"Thanks a lot." I mulled it all over. "What does this mean?"

"Well . . . taken with your little frustration yesterday, we can reasonably put together a picture. Unfortunately, it's a blurred picture."

"It's always been a blurred picture."

"Perhaps. But some parts of it are in focus now. For instance, Redwood's contention that the KGB was involved in Minnie's murder is more nonsensical now than it was when he first voiced it. In fact, what we know now suggests that the whole charade with the two manuscripts had no connection with her murder—except, of course, that she was murdered while in possession of the French manuscript."

"Then how can you say it had no connection?"

He put up a cautionary hand. "She was murdered *while* in

possession of the manuscript, however briefly . . . but not necessarily *because* she was in possession of it."

"That sounds like a pretty cockeyed distinction."

"I'm not entirely happy with it," he admitted. "But something tells me there is a distinction." He waved a hand at Franco "Let's order dinner while I mull that one over."

After ordering and mulling—he over whatever distinctions he had in mind, and I over the total muddle I was in—he seemed to get his presentation in order.

"Look at it this way," he said. "The charade proceeded as planned, except that one manuscript showed up in New York instead of Paris. We can eliminate the KGB, except as a deliberate red herring—or Redwood herring, if you will allow me a little joke—a misdirection orchestrated by Redwood for his own purposes. This leaves us with the intruder-murderer, which even our friend Harrington would acknowledge, except that the missing manuscript shows up in your office. This means . . . what? Let's take it in its simplest, most direct implications. It means that someone from the Press, other than Redwood, was at Minnie's apartment . . . before, after, or during the murder."

"Couldn't it mean . . ."

He stopped me. "That the murderer is the one who swiped the manuscript from Minnie's apartment and put it on your shelf? No. Or let's say, not necessarily. The implication goes only so far. Someone from the Press, and not Redwood, was in the apartment. I'm trying to separate the elements in this blasted confusion. There's the murder—one element. And there's the manuscript—another element. I'm trying to think of them separately."

"Why?"

"Because I don't see the connection, and until I do . . ." He turned up his empty hands. "Until I do, I prefer not to connect them."

"You make it sound reasonable."

He laughed. "You don't sound as if you really believe that. But accept it. Think of the manuscript . . . and think of your shelves."

"Yes. Why was the manuscript put there?"

"It was the perfect hiding place."

"The perfect hiding place?"

"Dear boy, consider the beauty of it, the cleverness. You have in your hands—what shall we call it?—a hot manuscript, a really hot manuscript. You can't hold on to it; you don't want to dump it. It's going to be valuable to you, perhaps... somehow, in some way. You can't hold it, but you want it where you can get your hands on it. There's your colleague, Miller. His shelves are loaded with a chaotic mess known as dead matter. And what is dead matter? Manuscripts, galleys, proofs... packages and packages and pages and pages of exactly what you've got burning your hands. Shove your hot manuscript in among his dead matter.

"Will Miller notice? Why should he? Your manuscript looks exactly like all the other dead matter. Will Miller look through his shelves? Why should he? He hasn't for ages. And what if he does? What if he finds the manuscript? That's the beautiful part. Confusion reigns, because Miller is the one person who couldn't have put the manuscript there. He was on a plane from Istanbul when the manuscript was removed from the apartment. That's why it's such a perfect place. Besides, I don't think it was meant to stay on the shelf for long."

"How do you know that?"

"I'm guessing that its value was rather immediate, or was considered immediate."

"How can you say that?"

"Precisely because it seems to have been abandoned."

"Now you've really got me confused."

"Well, what showed up that night? Or, from our unknown swiper's point of view, what showed up in the next day's papers?"

"The Russian manuscript."

"Exactly. When the existence of the Russian manuscript was revealed, the usefulness, whatever it was, of the French manuscript was negated. The obvious suggestion is that the swiper, presumably in possession of the one copy of an extremely valuable manuscript, was going to hold up our friend Redwood for a bit of money. But that's only a guess." He waved the guess aside impatiently. "In any event, the appearance of the Russian manuscript made the one on the shelf in your office useless."

"Are you sure?"

"Reasonably. Nobody reacted as I'd hoped when your shelves were emptied. I would say that's a good indication that the French manuscript was no longer useful. I still think emptying your shelves was a good idea. But obviously it didn't work. Because the usefulness of the manuscript was . . . over, gone, done with. It couldn't do whatever it was meant to do . . . so whoever put it there was glad to leave it. Hence, no reactions of shock, horror, surprise. No reaction but resignation."

I was trying to remember what everyone had said, how they had reacted. Resignation. Someone's phrase was coming back to me.

"It will come to you," Dobbs said. "It's rather obvious who swiped the manuscript and put it on your shelf. If you haven't figured it out by the time we have our *espresso*, I'll go back to the office with you, and we'll put into operation a little plan I have to prove who did it."

"Another plan?"

"This one will work," he said smugly.

We were back at the office as soon as dinner was over. I was as much in the dark as ever, but Dobbs was clearly pleased with himself. He wandered around my office, nodded appreciatively at the empty shelves, looked suspiciously at the armchair, and finally perched on my windowsill. I seated myself at my desk.

"Do you write interoffice notes or type them?" he asked.

"Write them." I was particularly proud of my handwriting.

"Ah, yes. I've seen that anal-erotic, compulsive handwriting of yours. Impressively clear. Anyone can read it." He picked up one of my memo pads, examined it, and handed it to me. "Here, this has your name on it. And the handwriting will make it doubly certain. Write what I say."

I looked over my shoulder at him.

"Trust me," he said. "Now let's see . . . okay, write this: 'I have the manuscript you left on my shelf. I will turn it over to the police today, unless you persuade me not to. I am at home.' That should do it."

I started to protest, but he stopped me. "Just write, It will all work out, have no fear." Then he dictated the message again, and I wrote it.

"I don't like this," I said when I was finished.

"Piece of cake," he replied. "You have nothing to worry about. All you have to do is call in sick tomorrow morning and stay home all day."

"What about the note?"

"Put it on Myra Palmer's desk where she's sure to see it first thing when she comes in in the morning."

"Myra?"

"But of course."

"Why Myra?" I was not as upset as I thought I'd be.

"Because the person who came to Minnie's apartment—before, during or after the murder—had to be someone from this office who was out of the office on the afternoon in question. Roger McGraw, Sherwood Leitner, Milt Foster, and Fred Snap were all in the office all afternoon. Harrington has assured me that the police statements confirm this. Redwood, of course, is covered, and you were on a plane from Istanbul. That leaves Paul Ostrow and Myra Palmer. And Paul knew about the Russian manuscript, which, as we know, eliminates him. It has to be Myra."

The way he said it, I couldn't argue. But I certainly felt uneasy.

"What if Myra really did it? What if she's a murderess? Why am I inviting her to come over to my apartment?"

"Dear boy, you really are a worrywart. Don't worry so much. You have my assurance that Myra Palmer is not a murderer. Why would she want to murder Minnie? The poor woman was her friend."

"That's no excuse. Maybe it was an accident."

"Maybe. But I sincerely doubt that. I told you, I prefer to think of the murder and the manuscript as separate."

"You prefer . . ."

"Oh, stop making a fuss. If it makes you feel any better, of course I'll be at the apartment with you. No doubt, Harrington and Preston will want to be there as well. Everything will be—"

"If you say 'piece of cake' once more, I'll hit you with one."

"You're overwrought. Now if you won't put that note on Myra's desk, I will." He reached for it.

"Never mind," I said, getting up. "I'll do it."

We left the note where Myra would be sure to see it, turned out the lights, and locked up the office. Down in the street, Dobbs was at his most cheerful.

"Let's take in a movie."

"I don't want to take in a movie."

"Don't be sulky. Perhaps you want to call your lady friend?"

"Not tonight. God, you've got my head so turned around, I've forgotten about her."

"Wonderful. It will do you both good."

Eventually he talked himself into staying over at his club and headed there. I took myself off to the subway. It was going to be a long night.

12

By nine o'clock, Dobbs, Harrington, and Preston were all at my apartment. There was an officious cheeriness about them that I found particularly aggravating. Dobbs kept smiling at me reassuringly. Harrington wandered around the room, touching the furniture and rearranging the throw pillows on the sofa. Preston kept telling me not to worry.

After about ten minutes of this, they excused themselves for a conference. I let them stay in the living room and went in and lay on the bed. From what little I could gather, the conference was about something they had found out about the French manuscript. Harrington had turned it over to Preston the first thing in the morning after we had found it. Since then, he had remained uncommunicative about it. Let them have their conference, I thought. To hell with them. I was annoyed that Dobbs was included, but not me. However, Dobbs had been helpful to Harrington, and possibly even to Preston, for all I knew. There was a lot about Dobbs, it seemed, that I didn't know.

They finished their conference and Dobbs poked his head into the bedroom. He had given up the irritating smile for a thoughtful look.

"Just getting filled in, dear boy. It's all very interesting. As soon as I make heads or tails of it, I'll give you a precis. If you're more comfortable lying down, why don't you just remain in here? We'll make ourselves comfortable in the living room. It shouldn't be long."

"Dobbs."

"Yes?"

"She's going to ring the bell. . . ."

"Or knock on the door. So what?"

"So what? So I go open it . . . and she shoots me dead."

"A picturesque but highly unlikely confrontation. Why should she shoot you on your opening the door?"

"So she can come in and get the manuscript."

"First of all, she wouldn't know where it is. Second, she doesn't really want the manuscript."

"Then why will she come?"

"To talk. I think she has a great need to talk. I don't know for sure what her motives were. But I am sure that her expectations have been terribly frustrated. From what you say, she's certainly unhappy. Oh, yes. She'll come just to talk. And we'll be happy to listen."

"Dobbs."

"Yes, dear boy?"

"How about if we leave the door open? And when she rings, or knocks, I call out 'Come in!' from the living room. That way she has to come all the way into the living room."

"Before she shoots, you mean? And you could be hiding behind a chair all the while. That way she could only kill me and Harrington before Preston, in an act of spontaneous bravery, wrests the gun from her."

I sighed and sat up. "It was just a suggestion."

"All right, we'll leave the door open."

I followed him into the living room. Harrington was sitting on the sofa, drumming his fingers against a pillow. Preston came out of the kitchen, eating a banana.

"I hope you don't mind," he said. "I didn't have any breakfast."

"I'll make some coffee," I said, "if you guys don't mind instant. I think I have enough doughnuts for all of us."

We all wound up preparing the coffee and doughnuts, to keep ourselves busy. The waiting was a strain.

It was close to ten o'clock when the phone rang. I looked at the others nervously, and Dobbs motioned me to answer it.

"Hello?"

"Howard, it's Myra."

"Yes?"

"I got your note. How long does it take to get to your apartment?"

"About half an hour. You take the—"

"I know how to get there. Just refresh my memory. What's the address and apartment number?"

I told her.

"I can't leave immediately." She sounded immeasurably weary. "But I want to come . . . I want to speak to you. I must speak to someone . . . I have to. . . . I'll be there about eleven." There was a long pause, then she added, "Howard . . . please don't go . . . anyplace. . . . Be there."

"I'll be here."

She hung up. I cradled the phone and looked at the others. "She'll be here about eleven," I said.

Harrington grunted. Dobbs rubbed his hands together. Preston stood up, then abruptly sat down again. No one seemed to have anything to say. I took myself off to the bathroom.

About a quarter to eleven, we arranged ourselves in chairs and on the sofa, in a semicircle facing the empty armchair in which I usually did my manuscript reading. Dobbs got up and fixed the blinds so that a shaft of light falling on the armchair was eliminated. Then he walked to the tiny entrance hall and opened the door. As he came back in and resumed his seat, Harrington asked, "What's all that for?"

"Just making it easier for her."

We waited.

Just before eleven, the sound of the elevator door and then footsteps came to us from the open front door. The doorbell rang, its sound unrealistically harsh in the quiet room. I jumped.

"Come in," I called.

She entered slowly, tentatively, carefully shutting the door behind her. When she walked into the living room, she stopped in surprise.

"Forgive us, madam," Dobbs said hurriedly, rising. "I realize you hadn't expected such a crowded reception. But please be assured it is a friendly reception. We are all your friends here."

Myra stood there, halfway into the room, blinking.

"It's all right, Myra," I said, also rising. "This is Hartley Dobbs. You know . . . 'Crime Cabinet.' He . . . he's the one who dictated the note I left for you." Dobbs stuck out his hand.

She ignored his outstretched hand and looked at the others. I, meanwhile, walked over to her and put a hand on her arm.

"It's all right," I repeated. "This is Mr. Harrington. And this is Lieutenant Preston. He's with the police."

She reacted slightly to that, but Dobbs and I guided her into the armchair. She sighed and even managed a smile as she sat down.

"You're all in on it?"

"Again, madam, forgive the deception," Dobbs said. " 'In on it' is not strictly true. The note was my device, mine alone. I was interested in smoking you out, so to speak. We tried it the other day by emptying Howard's shelves. That did not work . . . and I think I know why. No matter. This ruse has worked. I wanted to talk with you . . . or rather to listen to you, to hear what you have to say. Rather than have to repeat it all to these gentlemen—a tiresome prospect—I invited them to join us."

She let the smile linger a moment. "It's all right, Mr. Dobbs. You don't need to overdo it. I'm quite ready to confess—but only to filching the manuscript. I didn't kill Minnie."

"I know you didn't," Dobbs assured her. "Lieutenant Preston knows you didn't. Please be assured of that, madam."

She looked up at Dobbs with an expression almost of pleading, or perhaps it was anguish. "I'm glad," she said, "glad I took the manuscript, you know. I might have kept silent about it . . . even now. I hate Emory so. It serves him right. But . . . everything is turning out wrong. So I guess I'm better off telling you what happened. Isn't that so?"

I couldn't quite follow her, and I wondered if Dobbs could. Harrington was leaning forward on the sofa, and Preston, next to him, had taken out his notebook. Dobbs patted Myra's shoulder and went back to his chair.

"Yes," he said. "Tell us what happened."

Myra calmed down as she spoke, her tone becoming relaxed and conversational. The anguished confusion seemed to melt away.

She started out addressing herself to Dobbs; but soon she was talking to each of us, sort of holding court, like Gertrude Stein and a circle of admirers.

"When Minnie found that blasted manuscript in her portion of the slush," Myra said, "she called Emory. Though I suppose you know all that. Anyway, what you don't know is that she called me immediately after. She was in a terrible state. You people don't know what she was like . . . well, Howard can back me up. When Minnie got excited, she just . . . became incoherent. I could hardly understand what she was saying. Finally, I made it out. She had the missing manuscript, the one Emory had gone to Paris to get and make a big splash about and had come back without. Oh, he had made light of it . . . some minor mixup . . . but he was in such a nasty mood, even for him, we knew there had been a major foulup. We had been talking about it just that morning.

"So here she was, babbling away about having the manuscript. That's not very nice of me . . . I hope she'll forgive me . . . but it's true . . . she was babbling, and it was very difficult to understand her. She had the missing French manuscript, and she had called Emory, and he was coming over to get it, and she couldn't wait till he got there, and she was so nervous, and on and on. I was getting nervous myself, just listening to her."

"Excuse me, Mrs. Palmer," Preston interrupted. "Was there any particular reason why she was nervous? Did she say?"

"Not that I recall. You have to understand, Lieutenant, the conversation was hardly clear at the time, and this is days later. I can't really reconstruct it word for word. Besides, Minnie was the kind of person who would get in a nervous dither over a cake rising in the oven. Anyway, she was so upset I decided I would get over there at once, and told her so. She hung up and I started—"

"Another question." This time the interruption was from Dobbs. "I'm sorry for these interruptions, but surely you realize that when we ask you something, it's necessary. You're sure that Minnie asked you to come over and knew that you were on your way? She said something to that effect?"

Myra nodded. "Oh, yes. I said I was coming over and she said something like, 'Good, good, please hurry over.'"

Dobbs, Harrington, and Preston all looked at each other, and something like an un-nodded nod passed from one to the other.

"I started over at once," Myra continued, "and because I was so much closer, I arrived before Emory. I live only two blocks from Minnie's apartment. And let me tell you, I ran all the way. I didn't even put on a jacket. And I realized on the way that I still had my gloves on. When I read those slush manuscripts . . . some of them are so dusty . . . the envelopes they come in, from lying around so long . . . I usually put on an old pair of thin cotton gloves. Well, I was over there in a flash . . . and as soon as I got there, I knew something was wrong.

"For one thing, the door was open . . . not wide open, but unlocked and off the latch . . . slightly ajar. That wasn't like Minnie. She was always careful about locking her front door. So I just burst in, yelling, 'Minnie, Minnie, it's me.' I think I was calling to her as I came through the door and that little entrance hall." Myra stopped for a few seconds, the scene reshaping itself in her mind. "Anyhow . . . as soon as I stepped into the living room, I saw her. She was lying on the floor next to her work desk . . . and she was dead. I didn't know for sure, of course, but she had that look . . . sort of crumpled . . . and there was that big candlestick lying next to her. And the manuscripts were sort of strewn about on her desk . . . not like her, at all . . . and one of them just lying on top of her."

The three men glanced at each other. This time, Harrington raised a question.

"Excuse me," he asked, "a manuscript . . . on top of her?"

"Yes . . . as if she had been holding it when she fell."

"Now let me get this straight, Miss Palmer—"

"*Mrs*. Palmer. Though long since divorced, thank goodness. Surely you read your own police reports."

"I'm not with the police, Mrs. Palmer."

"Not with . . . ?" She eyed him warily. "Then . . . ?" She seemed to come to some conclusion that pleased her. "Aha!" she said with satisfaction.

"This is not entirely, or exclusively, a police matter."

Myra straightened up in her chair. "Yes," she said firmly, "a manuscript. But let me go on. It looked, even in that first instant, like a robbery, or an attempted robbery. There had been a number of robberies in the neighborhood . . . and the door was open. It looked like a robbery, and it looked like Minnie was dead. I bent down to examine her. But I panicked. I couldn't touch her. Besides, I was certain she was dead. I . . . I . . . should have called the police, I know. But all that was running through my head was that Minnie was dead and Emory was coming any minute. I wanted to get out of that apartment. It even crossed my mind that whoever had killed Minnie was still in the apartment."

Preston started to say something, but thought better of it.

"I was bent over her," Myra said, "and that manuscript lying on top of her was staring me in the face. It was half out of its envelope. I pulled it out farther, saw that it said something in French, and shoved it back in the envelope. I don't know exactly why, but . . . lying there on top of her crumpled body . . . that damned manuscript seemed to epitomize all my frustrations with everything at the Press . . . with Emory in particular. It struck me as his final indignity to poor Minnie, lying there. And then I saw it as a way of getting back at him . . . of somehow avenging Minnie. Poor Minnie, lying there so crumpled . . . as if she had been struck down by that manuscript . . . Emory's manuscript.

"I realized that I was wearing gloves. I had left no trace of being in the apartment. I had only been there a minute at the most. I scooped up the manuscript in the envelope and rushed like hell out of that apartment. I only stopped long enough to leave the door slightly open, just the way I had found it. And let me tell you, the thing that was uppermost in my mind was that Emory would be there any minute. I wanted him to come in and find Minnie. I wanted him to be stuck with her body. I was hoping somehow he might be blamed for her murder. Oh, don't ask me how—I don't know how. But I was hoping it."

She stopped to take a breath, and Preston quickly intervened.

"When you left the apartment," he asked, "how did you go back to your apartment?"

"What do you mean? I practically ran . . . if that's what you mean."

"No. What route did you take? Did you run down the block, or did you turn the corner? Minnie's apartment was on the ground floor of a corner building. Did you turn that corner?"

She thought for a moment. "Yes. That's odd. I turned the corner and hurried over to the next block, and then I turned the corner again and continued home. Ordinarily, I would have gone straight down the avenue and turned up at the far corner. It's more direct."

"Perhaps," Dobbs cut in, "you were unconsciously protecting yourself, avoiding the long avenue leading from Minnie's building, in case someone from the apartment came out after you—the murderer, for instance."

She shuddered. "I hadn't thought of that. But yes, it must have been some kind of instinct. I dashed out of the building and the street was empty . . . the long avenue block . . . it was so empty it looked surreal. I ducked around the corner. The side street looked—how should I say—safer. It was a short block. Then I turned the corner onto the next avenue. My building is there, two blocks down."

"Could someone have been following you?" Preston asked.

"I didn't see anyone."

"Did you look back?"

"Yes, as a matter of fact, I did. Two or three times. I didn't think of it until now. I guess I was pretty nervous."

"And if someone came out of the apartment," Dobbs said, half to himself and half to Preston, "his first choice would be to look down the open avenue block. Then, seeing no hurrying figure, he would try the corner. But by that time, Mrs. Palmer had turned the second corner and was hurrying down another avenue. Poor, perplexed pursuer."

"Lucky Myra," she added. "Do you really think there was someone in that apartment who came out after me?"

Preston looked at the others, then said, "It's a possibility. But," he said hastily, "please go on with your story."

She eyed him fishily, but continued. "Once I got home, I sat down with the manuscript in my hands and wondered what to do. I thought of calling the police—anonymously—and reporting the murder. I had a picture in my mind of policemen bursting into the apartment and grabbing Emory as he stood over the body. That was a satisfying vision. But I was afraid to call the police, because I had taken the manuscript."

"What did you hope to do with it?" Harrington asked.

"I don't know, really. Remember, I didn't know about *two* manuscripts at this time. All any of us knew about was this one important manuscript that Emory had gone over to Paris for . . . and hadn't gotten. Only now *I* had it. I had it, and he wanted it. That was all I knew, really. I was going to use it against him, somehow. Only I didn't know how.

"I kept turning it over in my hands . . . inside the envelope . . . I didn't even take it out and look at it. My French isn't that good, anyhow. I just kept fumbling with it. I didn't want it on me. But I didn't want to give it up, either. Then I had an idea. I decided to return to the office. Emory wouldn't be there. With any luck, he'd be in jail. I'd take the slush manuscripts back . . . there was nothing unusual about that . . . only I'd include the Kuzatov manuscript with them. So I did that. I went back to the office. A few people from the bullpen noticed me, but nobody paid particular attention. I dumped the slush in my office and took Kuzatov's manuscript into Howard's office. And I stuck it in amongst his piles of dead matter."

She smiled at me. "Nothing personal. It was just a good place to put it for safekeeping. You never touched the stuff. And the manuscript was invisible there—the old 'purloined letter' routine from Poe. Besides, if someone found it, you were safe. You were in Istanbul and couldn't have squirreled it away on your shelves."

I smiled at her halfheartedly and shrugged.

She shrugged back. "But nothing worked out right," she went on. "As soon as I read in the paper the next morning about the Russian manuscript, I knew I couldn't do anything with the French one. That was a pretty neat gimmick—the two manuscripts." She looked at Harrington. "Did you dream it up? Or was it Emory? It's certainly devious enough for him. Anyway, it threw all my plans into a cocked hat. Not that I really had plans . . . just wishes . . . and a lot of good they did me.

"Nothing was working out right, no matter what I wished. The French manuscript was useless. And not only wasn't Emory being charged with Minnie's murder—as if I thought that one would stick—but he was actually making hay with all his KGB nonsense." She banged her hand on the arm of the chair. "That's what really hurt. He was getting all that publicity . . . he was going to get rich. And why? Because I had run off with the goddam French manuscript. I had taken it to screw him . . . and instead . . ." She shook her head in bewilderment. "How could I have seen . . . ?"

"Nobody foresaw that one," Harrington intoned.

"All my schemes to get back at him," Myra mourned, "were just so much . . . dead matter. That's what they were, dead matter. Like the French manuscript. Like Minnie. Poor Minnie. Just so much dead matter. It's all a terrible mess. I'm glad to get it off my chest."

"You say you never opened the manuscript," Preston remarked.

"I looked at it for a second . . . in the apartment. It was in French. It was lying on top of Minnie . . . as if she had been holding it. I knew what it was. What else could it have been?"

"Oh, it was Kuzatov's manuscript," Harrington said. "You were right about that."

"A lot of good it did me."

"It threw a nice monkey wrench into things," he told her.

"Not for Emory. It worked right into his hands."

"When you picked it up," Dobbs asked her, "was it sealed or closed in any way?"

She thought about it. "No. The manuscript was half out of the envelope."

"There was a clasp on the back of the envelope," Dobbs suggested.

"The envelope was open," she insisted.

"Was there any tape on the back of the envelope? Think carefully, Mrs. Palmer. Do you remember seeing the little metal prongs? Were they covered?"

"I don't remember any tape. The clasp . . . I guess . . . was open. It must have been. The manuscript was half out of the envelope."

"There was no tape—like Scotch tape or masking tape—sticking to the envelope or hanging loose from it?"

"No. I'm sure I would have noticed it."

Dobbs looked inquiringly at Preston.

"We found no tape in the apartment," Preston told him.

Dobbs nodded, satisfied.

And that was that. The big confrontation was over. Myra sat disconsolately in the armchair. I looked at her blankly. Dobbs, Harrington, and Preston held a mumbled conference. The results of their mumbling seemed to please them. I managed to hear the word "tape," but little else. Whatever had satisfied Dobbs obviously satisfied the other two. I tried to think about it, but gave up, as much in the dark as ever.

Preston turned to Myra. "There are a few more questions," he told her. "And of course you'll have to make a statement. Do you mind coming down to the station with me?"

"Now?"

"Now will do just fine."

She nodded and got up out of the chair. Preston escorted her to the door. Everything about the way she moved suggested defeat. At the door, Preston turned to Harrington and Dobbs.

"We know where to reach you," Harrington said.

Preston nodded and left with Myra.

"Is there another phone?" Harrington asked.

"In the bedroom," I replied, with a wave in that direction.

He glanced meaningfully at Dobbs and went into the bedroom. Dobbs came over and put his hand on my shoulder.

"See," he said, "it all went very well."

"You're disappointed in me . . ." I murmured, "about . . . before."

"Dear boy," he said reassuringly, "you're just . . . well, imaginative. I'm sure it's a good sign in an editor."

"Thanks. I feel like an idiot."

"Just overactive imaginings. After all, from your point of view, she could have been a murderer."

"But she isn't."

"No, she isn't. Just meddlesome. I'm sure our friend Harrington is much relieved to know how the manuscript disappeared from Minnie's apartment."

"And what was all that about the tape?"

He smiled and ignored the question. "What you should

do now is go back to the office. Tell them you're miraculously cured."

"Back to the office?"

"Why not? We're done here. A little work at your desk will recharge your batteries. You look like your batteries need recharging."

"They need something. Where are you going?"

Again, he ignored the question. "I'll get in touch with you later. At least if you're at your desk, I'll know where to find you."

Harrington came out of the bedroom. "Ready?" he asked. "We're on."

Dobbs nodded. "Howard is going back to the office," he said.

"Good," Harrington said. "And not a word about this, any of it, to anyone, Miller. Remember that. Not a word of it to anyone . . . least of all to Redwood."

They left almost immediately. Harrington practically bubbled, but Dobbs seemed deeply thoughtful. I puttered around the apartment for a few minutes and then followed them out the door. On my way to the office, I picked up a sandwich and a chocolate milkshake. My stomach was in turmoil. For a cold, a Chinese meal; for a fluttery stomach, a chocolate milkshake. Those were my two steadiest home remedies.

It was lunchtime when I arrived at the office, and everyone was out. The emptiness seemed more unnatural than usual, but at first I couldn't figure out why. I was sitting at my desk, munching on my sandwich and slurping my milkshake, when I realized what was wrong. Mr. Snap was not at his desk. Every time I ate at my desk, I could be sure of one thing—Fred Snap would be next door, eating at his desk. His unexpected absence made me curious, if not uneasy.

I wandered out into the hall, double-checked Snap's empty office, and wandered down to the reception desk.

"Mr. Snap not in?" I asked the receptionist.

"Oh, he was here earlier, but he went out to lunch."

"Out to lunch?" I was flabbergasted.

She giggled. "Yeah, I know. Ain't like him at all. But I guess he's being taken." She paused and looked around. "That's the only way he'd ever go. It must have been that man that called him."

I looked at her inquiringly.

"Yeah," she said. "About half an hour or so ago, this man calls him—a Mr. Harrington. And then a little later, he comes out and says he's off to lunch."

I nodded my head in bewilderment and wandered slowly back to my desk. The chocolate milkshake did nothing for my stomach.

13 In the absence of any relief from my tried-and-true home remedy, I tried to soothe my stomach with a new possibility. I phoned Dinah. She answered on the first ring.

"Hello," she cooed musically.

"Hello yourself. I didn't expect to find you in. It's lunchtime."

"If you didn't expect to find me in, why did you call?"

"I wanted to see if you're still avoiding me."

"That's a good one. Who hasn't called whom for days?"

"I've been busy."

"A likely story. I've come to expect a daily call. Do you realize I'm building my life around your calls?"

That stopped me. I dropped my bantering tone. "Do you really mean that?" I asked.

"Of course not," she snapped. "I'm just mad at you for not calling me." Now *her* tone changed. "Oh, sweetie, it's the stupidest thing, and I can absolutely kick myself. But I missed you when you didn't call. Isn't that just awful?"

"It's wonderful," I said. "It's the nicest thing I've ever heard."

"That's because you're a sentimentalist."

"Say it again. Say you miss me."

"Why should I?"

"It makes my stomach feel better."

"What am I? A substitute for Maalox? Is that how you think of me? One teaspoonful whenever your tummy hurts?"

"You're better than Maalox."

"Gee, thanks."

"You're even better than chocolate milkshakes."

"That must be the nicest thing you ever said to a girl. Better than chocolate milkshakes. How do I rate against pizza pies?"

"Dinah . . . don't be mad at me. I'm in no condition. These last few days have been miserable. I need you to soothe me."

"Why? Have you run out of Maalox? Or milkshakes?"

"I'm full of both of them, as a matter of fact. I need you to run your lovely fingers over my fevered brow, and my poor, sick tummy, and my—"

"Never mind. I'm not interested in your anatomy. At least not on the telephone. Are you really sick?"

"Desperately."

"Good. It serves you right for deserting me. Are you in trouble?"

"I don't think so. At least not yet."

"You sound troubled. Are you going to tell me all about it?"

"How about tonight?"

"Suits me. I think I promised you a dinner, anyhow."

"Yes," I said eagerly. "This time at my place."

"Oh," she said archly. "That must be the key I got in the mail."

"You got the key."

"There seems to be an echo on this phone. Yes, I got the key. And I think I know how to use it. So don't hurry home. I'll get there early, anyhow. I have another screening to go to. I'll shop on the way down from the screening and have things started by the time you get there."

"How do you ever get any business done, with all those screenings?"

"Screenings *are* business," she said, and hung up.

I was feeling better already, and happily slurped my milkshake through the straw, making a noise on the bottom of the container like a motorcycle starting up. That sound always contents me as much as the milkshake itself.

In fact, I was contented enough to think about Snap lunching with Harrington without further upsetting my stomach. I was

197

wondering what they had to talk about, and whether Dobbs was with them. But my wondering produced no concrete results.

My thoughts were interrupted by Fran Bishop.

"I thought I heard you," she said from the doorway. "Didn't you call in sick this morning?"

"Yes. But now I'm feeling better."

"That's good." She shifted uneasily. "Emory didn't happen to call or come by while I was away from my desk?"

"Not that I know of. I've been here for about twenty minutes."

She peered down the hall in both directions. "I can't find him. I mean, he's not here." She fidgeted some more. "It's not like him."

"What's not like him?"

She seemed almost unable to bring herself to say it. "I just got a call from . . . his lunch appointment. Emory didn't show up at the restaurant. It's not like him to forget a lunch date."

"Maybe he's been delayed in traffic."

"No. He left quite early, much earlier than necessary. I thought maybe he had come back for something." She turned away in obvious distress, murmuring again, "It's not like him."

I gave it about ten seconds' thought and turned to the mail that had been left on my desk. I was reading letters when the phone rang. It was Harrington.

"How's your lunch?" I asked.

"Dobbs will tell you all about it. I haven't got time to talk now. I just called to ask if you happen to know whether Redwood's in the office. He was supposed to meet . . . an associate of mine, and he hasn't shown up. Is he there, do you know?"

"No, he isn't here. As a matter of fact, Fran Bishop was in here a minute ago, looking for him. She's rather upset because he skipped his lunch date. He left early, and she can't find him. She says it's not like him."

"It sure as hell isn't," Harrington agreed. "What's that son of a bitch up to? Oh, well, thanks, Miller. Listen, if Redwood shows up, let Dobbs know. He'll be able to get hold of me. Remember."

He hung up, and I cradled the phone in perplexity. What was going on?

I didn't have much time to puzzle over this latest development, because I could hear footsteps coming down the corridor and Dobbs's voice. He was explaining something to Mr. Snap.

They went into Snap's office, and a moment later Dobbs appeared in my doorway.

"Come join us," he said. "It's time for explanations."

Fred Snap was behind his desk, staring morosely at a pile of unopened mail. Dobbs was standing by the window, examining the dust-covered fan atop Snap's bookshelves. I entered and took a seat in the one armchair, an even lumpier replica of the one in my office. As soon as I sat down, Dobbs perched himself on the windowsill.

"Let us begin," he said, "with one thing that has remained constant in this shifting morass of confusions. The police, from the very beginning, have suspected an intruder of Minnie's murder. There have been, to be sure, numerous conjectures as to why the police stuck so emphatically to their intruder theory. These fancies notwithstanding, it turns out they have had, all along mind you, tangible evidence of an intruder's presence."

"Evidence?" I asked.

"Yes. The intruder conveniently dropped a glove in Minnie's bedroom. It is a man's glove and, as a matter of fact, a clumsy wool glove. Hardly the thing for purloining manuscripts, but perhaps satisfactory for bashing people with candlesticks. The police assumed it was clumsy; that it is wool is self-evident. Clumsy because our intruder took it off to open the envelope to make sure he had the right manuscript. This, once again, is an assumption, but a reasonably good one, because the intruder left a print on the manuscript.

"Now picture the scene as Lieutenant Preston reconstructs it. While Minnie is on the phone, our intruder comes up behind her; when she hangs up, he grabs her; at this point, perhaps in panic, perhaps by accident or intention, he bashes her, and lets her drop to the floor, a bulky envelope clutched in her hands. He bends over her, removes the envelope from her grasp, and attempts to extract the manuscript pages. Alas, he cannot. He is wearing woolen gloves, the last thing in the world you want to be wearing when you're trying to pick up sheets of paper. So he removes a glove, the left one, perhaps with his teeth, who knows? He gets one ungloved hand on the manuscript—thumb, upside down, clearly imprinted on top of page one—when he is interrupted. He hears a woman rushing through the hall toward the front door of the apartment. And the front door is open. And the woman is calling out to Minnie. It must be whoever she was speaking to only a moment ago on the telephone. What rotten luck! She has gotten here damned fast. The woman hurtles through the front door, and our intruder hurtles into the bedroom. There he drops the glove—perhaps from his chattering teeth. That's a dramatic touch of my own, not part of Preston's reconstruction. In any event, he drops the glove and provides the police with evidence of his presence in the apartment."

"Then he *was* in the apartment while Myra was there—hiding in the bedroom."

"Oh, yes. Consider the timing. It is more breakneck than in a Feydeau farce." Dobbs paused, then continued, "Remember Myra's description: the manuscript was lying half out of the envelope on top of Minnie's body. This is consistent with our picture of the intruder's movements. Myra swipes the manuscript and hurries out.

"Here we have to conjecture a little, but the comparison with a French farce remains apt. The intruder steals out of the bedroom. He sees no woman—no living woman, that is. She's gone. What's more, the manuscript is gone. He panics. We assume this because he is wearing one glove and has dropped the other, but doesn't seem to have realized it. He leaves by the front door, presumably in pursuit of the woman who has left with the manuscript. But he, too, pauses to leave the door slightly ajar. Then he rushes out to the street. There is no one to be seen. He runs to the corner. But too late—Myra has had time to turn the second corner. Again, there is no one to be seen. Now he really panics. He cannot return to the apartment. Presumably he has heard both of Minnie's telephone conversations, so he knows that Redwood is bound to arrive any minute. He flees. And shortly after, Redwood arrives at the apartment. Split-second timing and outrageous coincidences that Feydeau could not have contrived any better."

"Only Feydeau wouldn't have included a murder."

"True," Dobbs agreed, "but there is always the possibility that the murder was accidental. Life is an imperfect imitator of art."

"How much of this is certain?" I asked. "And how much is just conjecture?"

"Oh, it's pretty certain . . . considering the subsequent developments."

"There's more?"

"Much more, dear boy. When I promise explanations, I don't mean half-told stories."

"Oh, there's plenty more," Snap growled.

Dobbs got up from his perch on the windowsill and began pacing as he continued with his account.

"Now skip to last Friday," he said. "The police are called to a cheap rooming house on the Lower East Side. There, a man lies dead—an old drunk, as he is described by his landlady—apparently a suicide. There is a plastic bag over his head, and he is lying fully clothed on his bed. It is not an uncommon method of suicide.

"The police arrive, and even a summary examination at the scene reveals some unpleasant inconsistencies. There are slight signs of pressure applied to his throat, indicative of throttling, though perhaps not strong enough to indicate strangulation. Then too, there are signs of pummeling. Again, not forceful, let alone lethal, bruises, but indications of some kind of manhandling. The poor old man may have been forced to put the plastic bag over his head, or he may simply have been roughed up in an argument prior to his unfortunate demise."

As he eased into this narrative, Dobbs assumed precisely the tone of his television persona. He glanced occasionally at me or at Snap, almost as if he were looking for the little red light that indicated which camera was "live" at the moment. I was amused, but also, as with watching him on television, I was captivated.

"The police began to think of murder," he continued. "But who wanted to murder this sort of fellow? A drunken derelict—an old man and, from all appearances, nearly penniless. Suicide was not inconsistent with his condition, someone near the end of his rope. But there were those signs of manhandling. So the police investigated.

"First of all, they checked out the fellow's identity, which didn't present any problem; all sorts of people in the building could identify him. His name was Bruno Korzenyi."

Snap grunted or snorted and shook his head.

I was startled. There was something familiar about the name.

"You look puzzled," Dobbs said to me. "Perhaps it will come to you. Mr. Snap, of course, is quite familiar with Korzenyi. But let me continue. Bruno Korzenyi . . . a European refugee from way back. The other lodgers and the landlady were voluble enough. The police came away from the rooming house with a notebook of material, some of it even helpful. They had enough information, or leads, to suggest one immediate course of action—they checked Bruno Korzenyi through Washington.

"Now, for the moment, we leave this possible suicide, possible murder. It is Friday afternoon. The body has been removed. Its identity has been established. The vast bureaucratic machinery of Washington has been set in motion.

"As of Friday afternoon, all this is merely a digression. Lieutenant Preston, who is not involved, and who is busy with his own concerns, does not even know of the existence of Bruno Korzenyi. Myles Harrington, who does know of Korzenyi, has other things on his mind, and does not even think of the man or know of his sudden death. Howard Miller is on a bus to New Jersey."

I looked at him inquiringly.

"A mere dramatic flourish," he said apologetically. "A device to maintain your attention. We skip now to Tuesday of this week."

I leaned back in my chair. Snap settled noisily in his.

"On Tuesday morning," Dobbs continued, "you will recall, our friend Harrington visited Lieutenant Preston with the newly found French Kuzatov manuscript. The police laboratory went to work on it immediately, and in short order produced an unidentified print—thumb, upside down, on the top page. We have already discussed how and when it got there. But wait! Is it really unidentifiable? Lieutenant Preston, whose visual memory is truly remarkable, says it strikes him as familiar . . . that scar crossing the whorls. . . . He hunts on his desk for a report that he had been reading when Harrington arrived—a report from another precinct about a suspicious death. Need I belabor the obvious, the inevitable? The thumbprint on the Kuzatov manuscript is Korzenyi's."

"What a coincidence," I muttered.

"A coincidence? Oh, yes, perhaps. But merely a time-saver. In the course of that day or the next, bureaucratic procedure would have produced the same result. The print would have been identified soon enough."

"So Bruno Korzenyi was the intruder," Snap stated. "Did he murder Minnie?"

Dobbs held up a cautioning hand. He was going to tell this his way. I could see the television showman at work.

"You can imagine Harrington's reaction. He knew exactly who Korzenyi was. He knew all about him. Korzenyi had done work for Harrington in the past. But that had been a long time ago. Korzenyi, it seems, had been an old drunk for years before his untimely shuffling off. The CIA had had no use for him for some time. And so poor old Bruno had been existing on handouts, had been sponging off his friends. And one of his principal benefactors—this is what had Harrington so excited—"

"Emory," Snap intoned gloomily.

"Precisely! One of Bruno Korzenyi's last contacts with the old days, one of his few friends and benefactors, was Emory Redwood."

Dobbs paused dramatically. I had the sense of a noose being tightened, but around whom or what I couldn't quite figure out. Dobbs's "showbiz" delivery was getting to me.

"So we can reasonably suppose," said Dobbs, "that Korzenyi was hired to steal the manuscript. It is not a reasonable assumption to place him in Minnie's apartment of his own volition. But that presupposes two things: one, that someone hired him; and two, that the person or persons who hired him knew that the missing manuscript would be in Minnie's apartment."

"Was he supposed to murder Minnie?" I asked.

"That could be. I'm not certain. As I said before, the murder could have been an accident. But if you will follow me as we explore the motive for the stealing of the manuscript—for its being in Minnie's apartment in the first place—you will find, unfortunately, that Minnie's murder fits very consistently into the plan."

He stopped pacing and resumed his perch on the windowsill.

"So Korzenyi was hired to steal the manuscript from Minnie's apartment and maybe even murder Minnie—certainly attack her, that much is painfully clear. But by whom? The KGB? No, that's nonsensical. We have buried that smelly red herring, I should think. No. It's much more likely that the person who hired Korzenyi was Redwood."

"Emory?" I burst out. "Why Emory?"

He was waiting for that. "You will recall," he said in his best television manner, "the essential clue I pointed out in your office Monday evening—the clue I referred to again, when we questioned Mrs. Palmer this morning. It directs us irrevocably to Redwood."

I looked puzzled. He turned from me and directed his explanation to Mr. Snap, who stared glumly at his desktop.

"When I first examined the envelope in which the French Kuzatov manuscript arrived, Monday evening, next door in Howard's office, I noticed that it was not only unclasped, but unsealed. Nevertheless, it was apparent from even a brief examination that it had originally been sealed. When it was mailed, the flap had been closed and tape of some sort had been fixed over the prongs to make it more secure. Think of mailing a manuscript—an important manuscript—overseas. It stands to reason that the mailer is going to secure the envelope with tape. In any event, I could see where the tape had been removed; we could all see where the tape had been removed.

"But there was no tape on the envelope when Mrs. Palmer found it on Minnie's body. And there was no tape lying on the floor or anywhere in Minnie's apartment. Lieutenant Preston confirmed that. What does that mean? It means that the envelope was given to Minnie minus the tape. And what does *that* mean? It means that the envelope had been opened earlier, before it was ever handed to Minnie."

"When?" I asked.

"First ask: Where? And the 'when' will follow," Dobbs said.

"All right, where?"

"In the offices of Redwood Press, of course. It's the only logical place. Don't you see yet how neatly it all dovetails?"

Snap grunted some kind of acknowledgment.

"Who handed out the slush manuscripts?" Dobbs asked patiently.

"Emory."

"And so, who saw to it that Minnie got Kuzatov's manuscript? Who made absolutely certain that she got it?"

"Well . . . Emory. But he could have been just handing them out haphazardly . . . and Minnie happened to get it."

"Even though he knew perfectly well it was in the pile? Hand them out haphazardly? Not . . . bloody . . . likely."

"Why Minnie? Why did he have to give it to her?"

"Because he could count on her response. Even if she didn't phone him in her excitement, he could have found some excuse to go to her apartment and find the body. All he had to do was *say* she phoned him. But he was pretty certain of her response. And of course he was right. The only thing he didn't know was that she had also phoned Myra Palmer."

"Myra . . ." I said. "Why didn't he give it to Myra?"

"I considered that," Dobbs said. "You must understand that the one who got the manuscript was going to be a victim—I'll get to that in a moment—and he may have been very happy with the thought of making Mrs. Palmer a victim. There was certainly no love lost between them.

"But he had to choose the one of them whose response he could count on. As we know from subsequent developments, he was very wise in not giving the Kuzatov manuscript to Mrs. Palmer. I think he understood that instinctively, and ruled her out."

"What did you mean about a victim?"

"Minnie was not only supposed to be robbed of the manuscript, she was to be forcibly robbed—bopped on the head, possibly even murdered."

"But why? And when did Emory have time to plan all this?"

"The night before. That's almost certainly when he found the manuscript in the slush pile . . . when he was alone in the office, fuming about the mess-up of his carefully planned operation.

"Consider the psychology—not merely Redwood's, but the whole background of publishing. This publishing business is essential to an understanding of what happened. There's Redwood, fuming alone in the office, probably tearing the place apart in his frustration. You said that Harrington told you Redwood wanted to fly out to Istanbul. But he doesn't; he stays here, stalking around the office like a wounded tiger. He comes upon the slush pile . . . it's enormous . . . another frustration, his editors aren't doing their jobs . . . and then, lo and behold! What's this? The missing French manuscript.

"He must have been furious. But he was also a publisher—someone who had spent his entire adult life in the business. He accepted the fact that the missing manuscript had somehow turned up in the slush pile." Here, Dobbs smiled maliciously at me. "Remember how that bothered Harrington, but didn't faze you a bit? You publishing people accept things like that—manuscripts misplaced in slush piles, hidden away among dead matter. It gives reasonable people pause. It raises the hairs on the back of an author's neck. But it doesn't faze you publishing types."

"You said a mouthful," Snap put in.

"But to continue. Redwood had undoubtedly planned all sorts of publicity around the two manuscripts. He wanted a big news story to help publicize the Kuzatov manuscript. Now he saw an even better story. The plan must have formed in his mind almost at once. Sometime that evening he got hold of his rummy old cohort Bruno Korzenyi—who was deeply indebted to him, who would presumably do anything for him . . . and for money. 'Look,' he says to Bruno. 'Go to this apartment at precisely such and such time; force your way in and swipe this particular manuscript; maybe even bop the lady.' Oh, it may have taken a little explaining, but it probably didn't take much persuading.

"The details will need some sorting out. But it is entirely consistent with what we know to suppose that Redwood had it all worked out that night. Minnie would get the manuscript. Bruno would take it away from her. And the whole big KGB fuss would be put in motion. And *that* would really be a big news story—better publicity for the book than he ever could have hoped for from the original scenario."

"I can't believe this," I said, knowing that, in fact, I did believe what Dobbs was saying. "Why would Emory do all that?"

Dobbs considered his answer. He glanced at Snap, who performed a kind of combination shrug and sigh, while still staring at his desktop.

"Desperation," Dobbs said.

Snap raised his head and nodded sadly in agreement.

"The Kuzatov book," explained Dobbs, "would be worth considerably more money in sales and in subsidiary rights if there was enough publicity attached to it from the beginning. That was part of the original scheme—the two manuscripts. In a sense, two manuscripts was simply a precautionary measure. But Redwood dreamed up the business of publicizing his trip to Paris while you went to Istanbul undercover. He was building up a big story to publicize the book and make it more attractive for book club and paperback deals. When the situation changed, he saw an opportunity for even more publicity. So he took it."

"Why was he so desperate for publicity?"

Again, Dobbs glanced at Snap.

"The Press desperately needed the money," Snap said tonelessly. "We were going under unless we could come up with a very big deal. Emory was convinced that he could generate two or three hundred thousand at least, in paperback money, with the proper publicity. He talked Harrington into letting us arrange for the delivery of two manuscripts, the one in French in Paris, the Russian one in Istanbul, to help publicize the book. He didn't tell Harrington how close we were to going bust. He just convinced him that the publicity would help sell the book and spread Kuzatov's message."

"You can understand Harrington's annoyance," Dobbs interjected, "when Redwood started his KGB nonsense. Harrington's concern was in getting the book published here. I daresay he really didn't care whether it was translated from the French or the Russian. To him, Kuzatov's message was the important thing. When Redwood generated a whole circus around the book, Harrington was furious. Eventually he decided publicity was publicity, and if people bought the book for the wrong reason, but in greater numbers, that was all right."

"Are we really going bust?" I asked Snap.

"You keep that under your hat," he growled. "Besides, we've been reprieved. Emory got over six hundred thousand for that damned book." His eyes suddenly took on a dreamy look. He came as close to a smile as he ever allowed himself. "And with the new publicity that's going to break," he mused, "Redwood Press is going to be hotter than ever. I'll probably be able to sell out to a conglomerate."

Dobbs and I stared silently at each other, contemplating the ways of the world.

"Yes," Dobbs said, breaking his reverie, "the publicity. You people had better prepare yourselves." A thought suddenly occurred to him. "Good heavens, so should I! I'm a Redwood Press author. Do you think this house can exist without Redwood at the helm?"

"What are you talking about?" I asked fearfully.

"Face it, dear boy. What do you think happened? Redwood went through his big KGB act thinking Korzenyi had taken the manuscript from Minnie's apartment. When he finally caught up with the old man, there was, of course, no manuscript, and . . . well, we know what happened."

"Emory killed him?"

"He must have thought the old boy was holding out on him. There's the evidence of Korzenyi having been roughed up. Yes, I'm sure he killed him, even if it was by accident. No doubt, Redwood will say it was an accident. But it has to have been Redwood who placed a plastic bag over the man's head. That's hardly accidental. No. I'm afraid there is no way of keeping the esteemed leader of Redwood Press out of the pokey. You're going to have to get along without him."

"We'll manage," Snap stated.

"I can't think of Emory as a murderer," I said.

"Think of his state of mind," Dobbs told me. "He still doesn't know where the missing manuscript is. By the way, where is Redwood?"

"Oh, I forgot to tell you. He's missing."

"Missing? What are you talking about?"

"He was supposed to have lunch with one of Harrington's buddies, but he never showed up. Fran Bishop is worried about him."

"She's not the only one. Does Harrington know?"

"Yes. He called up here to ask me if I knew where Emory was. He said if Emory showed up, you were to let him know immediately."

"How can I let him know if I don't know myself?" Dobbs jumped up and resumed pacing. He stopped in the doorway, put his hand to his chin, and began staring thoughtfully in Snap's general direction.

"Don't look at me," Snap muttered. "I don't know where he is."

"No, of course not," Dobbs said in a faraway mumble. He continued to contemplate some point beyond Snap's head. "Nor do I. But," he said, suddenly brightening, "I have a damned good idea where he's going to be later in the day."

"Where?" we asked in unison.

"Gentlemen, you'll have to excuse me," he said briskly. "I'm off to demonstrate that I haven't lost my touch—I hope."

He dashed out, leaving us to stare at one another.

About an hour later, after a futile attempt to do some work at my desk, I wandered down the hall to Paul Ostrow's office. He wasn't there. Sherwood, coming out of his office, saw me looking in at Paul's empty cubicle.

"He's not there," Sherwood said. "He got a call from the police. All very mysterious."

"Mysterious?"

"He stopped by to tell me they'd called him. They wanted him to come down right away to the station house. but they wouldn't say what it was about. I call that mysterious."

I shrugged and went back to my office. I sat at my desk and tried to puzzle out the latest developments. Why had Dobbs dashed out so precipitously? And where had he gone? Where was Emory? Where did Dobbs think he was going to be? And why had the police summoned Paul? I was getting nowhere, when I realized someone was standing in my doorway. It was Fran Bishop, with a puzzled frown on her face.

"It's Emory," she said in a subdued voice. "He's on the phone. He wants to speak to you."

"To me?"

"Yes. He asked for Paul first, but I told him he isn't here. Then he asked for you."

"Did you tell him Paul was with the police?"

She looked startled. "I didn't know that."

"Never mind," I said, getting up from the desk. "Where's Emory, on your phone?"

"No, on his. In his office." As she followed me down the hall, she touched my arm tentatively. "I hope he's all right," she said. "He sounds so peculiar. I don't like it."

You're going to like it less, I thought bitterly. She hovered in the doorway while I crossed to Emory's desk. Then, when I picked up the phone, she discreetly went back to her own office.

"Hello, Emory," I said cheerfully.

"Hello, Howard," he responded in his best Viennese-dentist voice. "You people all seem to be having such a busy day without me. I'm glad to see you returned to the office. Was it an interesting morning at home?"

"I don't know what you mean."

"Howard, let's stop playing games. Myra left your cryptic note on her desk when she departed so unexpectedly this morning. I happened to see it when I went into her office. It puzzled me at first. But I can put two and two together. Is she the one who removed my manuscript from Minnie's apartment?"

"Uh . . . she's not here now."

"I didn't ask if she's there. I asked if . . . oh, never mind. Your fumbling answer is good enough. So she took it. She must have gotten to the apartment ahead of me. Bruno kept insisting a woman was there. So she took it. Damn her interference. So the police probably know. Did you have the police at your apartment this morning? Or is that where she is now? Do the police know?" His voice hardened. "Do they?"

"Emory . . . I really . . ."

"Ah, so. Tell me, Howard, do they also know about Bruno Korzenyi?"

"Really, Emory . . . I . . . I shouldn't . . ."

"Oh, stop *phumphing!*" He paused, and then his voice became more conciliatory. "Never mind, Howard. You don't need to answer. You never could lie properly. That's one of your more endearing traits, however exasperating it can be at times."

"Emory, where are you?"

"Where? That's for me to know." Again his tone changed. He was suddenly all business. "Howard, listen! I have no time for games. As you can well imagine, I have had to alter my plans unexpectedly. I am in a very delicate situation. Things have not gone . . . they have gone against me. I'm sure I must look very bad to you at the moment. But believe me, Howard, things are not as they seem. I can explain everything to you. You have to see my side of it."

"Your side of what? Minnie? Bruno Korzenyi?"

"Howard, Howard, I must explain. I can explain it all to you. I assure you, you will understand. You will see I am not so terrible."

"Why don't you explain to the police?"

I could hear the hiss of his breath, then a sighing murmur that sounded like, "Howard, Howard."

"Talk to Harrington."

"Don't be preposterous. He hates me. Howard, listen to me. You are the only person I can turn to. As it happens, I am perfectly confident of getting to . . . someplace where I can tell my side of the story. What I need is a little time and a certain amount of assistance."

"Assistance? You expect me to assist you?"

"Yes! I expect you to assist me." His tone turned icy. "We have worked together for more than three years. In that time we have come to understand each other. Just recently, I put my entire trust and confidence in you. Now I am asking you to trust me. I think you owe me that much. That trust is something we owe each other . . . something we share. Yes, I expect you to assist me."

I crumbled. "I won't break the law, Emory."

"I am not talking about the law. Forget the law—ignore it. I am looking for time, a little assistance, and silence. I need time, Howard, and a little of your help . . . and above all, silence. You must not say a word of this conversation to anyone—not to the police, not to Harrington—not to anyone."

"But Emory . . ."

"I will see you in . . . let's see . . . what time is it?"

"About four-thirty," I said, glancing at the clock on his desk. "But—"

"That's good enough. I give you permission to leave early. How long will it take you to get home? I'll meet you at your apartment at five."

"My apartment?"

"Certainly. It's the best place. Hurry! Leave now."

He hung up, and I stared helplessly at the phone in my hand. What had I gotten myself into? Why did I always let Emory twist me like a pretzel? My first impulse was to call Dobbs for advice. But I didn't even know where he was, since he had dashed off so precipitately.

I cradled the phone and headed for my office. Fran Bishop was waiting for me in Emory's doorway. I was so distracted, I almost bumped into her.

"Is he all right?" she asked. "Where is he?"

"He's all right," I said reassuringly. "I'm going to meet him."

"Where is he? Does he need help? Can I do anything?"

"It's all right, Fran. I'll take care of it."

"Bless you," she said, looking as if she were going to cry.

I brushed past her, wondering if blessings were appropriate for what I was doing, and went into my office for my jacket. I left the building and headed for the subway. Something was nagging at me, but I couldn't think what it was. I was standing on the subway platform when, in a rush of horror, it occurred to me.

Dinah was at the apartment!

She had said she was going to a screening and then was heading straight for the apartment. She had a key. She would be there when Emory arrived—unless I got there first. I looked down the tracks in an agony of anticipation. There was no train in sight.

14

That was the longest subway ride I ever took, even though it actually amounted to only a matter of minutes. I thought of dashing out of the subway station and taking a taxi. But, traffic being what it was at that hour, I knew I would get uptown faster underground. If only the train would come. I fretted impatiently while waiting for it to appear, and when it finally arrived, I plunged through the doors before they were fully open. When I got to my station, I dashed out of the train and up the steps, and ran all the way to my building.

As I opened the door to my apartment, I could see into the living room. Dinah was seated in a chair near the window. She was holding herself in a tense and hostile way, and glaring fiercely at someone out of my line of vision.

"Dinah," I called, slamming the door behind me. "Are you all right?"

She transferred the glare to me, as I came charging into the room.

"I told you he was a nut," she said severely. "And now he's gone right off the deep end. He's threatening me."

"Not threatening," Emory said, "just waiting with you. And I am not a nut, Miss Foxworth, just cautious."

"Cautious," she sneered.

I stood there looking at them, breathing heavily from all my running, unable to speak and not sure, anyhow, of what to say.

"Howard," Emory said in a placating tone, "imagine my surprise when I knocked on your door and it was opened by the esteemed Miss Dinah Foxworth." He raised his eyebrows. "She seems to have been cooking dinner for you."

"Does he run everybody's private life, too?" Dinah asked angrily. "Let me know when he shows up for bed check."

"You didn't tell me Miss Foxworth would be here. I'm a bit caught off guard." He smiled his dentist smile. "And impressed."

"Quick, Howard," Dinah cut in, "ask him for a raise."

Emory turned to glare at her. I finally found my voice.

"You didn't give me a chance, Emory. It wasn't my idea to meet here. But now that we're here, why don't you and I go someplace to . . . conduct our business . . . and leave Dinah alone to cook dinner?"

"Don't be fatuous," he snapped. "Now that she's here, you'll both have to help me."

"Leave her out of it, Emory."

He looked at me craftily, and then at Dinah. "No, Howard, I think that Miss Foxworth's fortuitous presence will ensure that you do your utmost on my behalf."

"What's going on?" she asked, her belligerence suddenly gone.

"Emory's in trouble with the police," I told her.

"The police? What did he do?"

"Oh, nothing much; just arranged one murder and possibly committed another. Isn't that right, Emory?"

"I told you, I can explain everything."

Dinah gasped and started to get out of the chair. Emory moved over to her swiftly and tried to push her back.

"Leave her alone!" I was very angry now, and my voice showed it. I stepped forward menacingly. "Get your hands off her, or I'll break every bone in your goddam body."

Emory was startled. He backed away from the chair and flashed me a nervous smile. "Howard," he said hurriedly, "don't be silly. I didn't mean her any harm." He continued backing away.

"You'd better not." I advanced toward him a few steps, and reached out my arm for Dinah. She scooted out of the chair and stood behind me.

"All right, Emory," I said gruffly. "I'll listen to your explanation. Start talking."

He blinked a few times and tried another smile. "Howard, what has come over you? We're all reasonable people." A little assurance returned to his voice. "Of course, I'll explain."

I stood there looking at him. Dinah slipped her hand in mine.

"Start with the French manuscript," I said. "You found it in the slush pile that Monday night, when you were alone in the office."

Emory's eyebrows went up. He hesitated, and I swear I could see him visibly rearranging his thoughts. "You seem to be . . . well informed. Perhaps the police . . . but no." He frowned and shook his head.

"We're all better informed than you think, Emory. You'd better just give it to me straight."

"Of course," he said agreeably. "Yes, I found the blasted manuscript in the slush pile. How it got there, I don't know. But it was obvious in Paris that something had gone wrong. As soon as I saw it, I knew I had to rearrange my plans. Of course, there was still the Russian manuscript. But I had no fears about that one. I knew I could count on you." He gave me the dentist smile again. "I have always been able to count on you."

I stared at him impassively, and he dropped the smile.

"Well," he continued, "at once a new plan presented itself. You understand, maximizing the publicity value of the two manuscripts was always at the bottom of my figuring. I concocted the business of making a fuss over the slush pile the next day—so I could hand the Kuzatov manuscript over to Minnie Heffernan. I was pretty certain how she would . . . well, that she was the one to give it to. Then I met with an old colleague of mine, to arrange for him to . . . appropriate the manuscript from Minnie. I knew I could rely on him. He was . . . well . . ."

"An old drunk, deeply indebted to you," I put in. "Yes. I'm sure you could rely on Bruno Korzenyi. Was he supposed to murder Minnie?"

"Of course not," Emory expostulated. "That was entirely an accident. He was merely supposed to threaten her and grab the manuscript." He paused to see how this was going down. "Well . . . maybe rough her up a bit," he added lamely. "I didn't expect him to hit her with that candlestick. Of course, when I arrived at Minnie's apartment and saw her body . . . I knew who had done it."

"And you assumed he had gotten away with the manuscript. So you launched into that whole KGB circus, knowing all the while that Bruno was the one who had killed Minnie."

"I was interested only in maximizing the publicity potential of the manuscript," he protested. "And of course, by deflecting attention to the KGB, I was covering for Bruno."

"Of course," I sneered, "protecting Bruno, looking out for his welfare. Were you looking out for his welfare when you put a plastic bag over his head? Or did you just go berserk when he didn't have the manuscript?"

Emory paled and licked his lips. "You have it all wrong."

"Do I? How do you explain Bruno?"

"Well, of course, I was upset. He kept insisting he didn't have the manuscript. And I was positive he did—that he was just holding out on me for more money. How was I to know that damned interfering woman had taken the manuscript?"

"Didn't Bruno try to tell you?"

"He wasn't making sense . . . going on about some woman coming in and surprising him. It didn't sound likely. Besides, an old drunk . . . a devious and greedy . . . old . . ."

"So you beat him up."

"I didn't beat him up," Emory shouted. "I pushed him a little."

"And he dropped dead of a heart attack."

"Yes, yes," Emory agreed eagerly. "Suddenly he was lying there."

"With a plastic bag over his head."

"I found that in the apartment. It was a bread bag. I put it over his head and placed him on the bed, to make it look like suicide. I thought if it looked like . . . I mean . . . after all, his dying like that . . . I didn't . . . it was an accident. . . ."

All the punch went out of Emory. He stopped talking, sighed, walked over to the armchair, and sat down heavily. I moved over to the window, keeping an eye on him. Dinah clung to my hand.

"We have to tell the police all this," she said.

"The police already know," I told her.

Emory looked up. "What do they know?" he asked. "What do they have that they can hold against me?"

"A little piece of tape," came a voice from the bedroom doorway. "A little piece of tape that they're going to make stick."

Hartley Dobbs came into the room, smiling benignly.

"Who is that?" Emory shouted, jumping up.

Dinah shrieked and ducked behind me.

"Who is that?" Emory shouted again.

Dobbs smiled at us, turning from one to the other. He looked as if he were waiting for applause.

"Emory," I said, "meet Hartley Dobbs."

Dobbs nodded and turned to Dinah.

"Dinah Foxworth," I went on agreeably, "Hartley Dobbs."

"Christ!" she said from behind me. "Do you always make entrances like that?"

"How did he get in here?" Emory demanded.

"With the help of the police," Dobbs said pleasantly. He turned to me. "You really should get a more secure lock, dear boy. That's a professional opinion from the officer who let me in."

"What are you doing here?" Emory demanded.

"I believe it's called eavesdropping. But it was very instructive. You realize, Mr. Redwood, that three persons have overheard your . . . conversation. Dare I call it a confession?"

Emory stared at him unbelievingly for a moment, then dashed for the front door. I started after him, but Dobbs held up a restraining hand. Emory flung the door open and stopped short. Two bulky figures filled the doorway—Harrington and Preston.

"Hello, Redwood," Harrington said pleasantly. "We've been looking all over town for you."

Emory backed into the room. He turned angrily toward me.

"Don't blame Howard," Dobbs told him. "This entire rendezvous is my doing. I invited Harrington and Preston. I ensconced myself in the bedroom earlier, without Howard knowing a thing about it. You see, if you were relying on Howard, so was I. I was relying on his loyalty to you . . . just as I was relying on your willingness to use people." He nodded his head in satisfaction. "It's so gratifying when people act predictably."

"Well, everything worked out just as you planned," Harrington told him with evident appreciation.

"Almost as I planned," Dobbs replied. "I must admit, I was momentarily thrown for a loop when a young lady arrived . . . with her own key . . . and began preparing dinner."

Dinah giggled. "Were you in the bedroom all that time?"

"Yes, my dear. But I was worried when Redwood knocked on the door and you so blithely went to open it."

"I thought it was Howard."

"Oh, well," he said, "it all turned out all right in the end."

"Thanks to you," Harrington said.

"And thanks to Howard. You should have heard him, Myles. He handled Redwood just perfectly." Dobbs beamed at me. "First he threatened to break every bone in Redwood's body, then he very skillfully led him into very damaging admissions."

"Did Redwood confess under physical pressure?" Preston asked anxiously.

"No pressure or coercion of any kind," Dobbs assured him. "On the other hand, Lieutenant, not much of a confession. Of course, what we'll testify to hearing is damaging enough. Redwood will need a good lawyer."

Emory made a growling, sneering noise.

"Don't snarl at me, Redwood. Our statements as to what we heard here this afternoon will only be the icing on the cake. Yes. You shall need a very good lawyer indeed."

Emory turned to Preston, who was hovering over him. "If you intend to take me someplace, then do it."

"I'll be down in my own car in a few minutes," Harrington said to Preston. "I just want to clear up a few things with these people."

Preston nodded and ushered Emory out of the apartment. As he opened and closed the door, I could see uniforms out in the hall.

"I think maybe you're a genius," Harrington told Dobbs. "Or are you going to tell me you were just lucky?"

"Oh, luck had a part in it," Dobbs replied modestly. "But I was, as I said to Redwood, counting on the predictability of people."

"How did you know he was going to come here?" I asked.

"He was going to turn to you or Paul Ostrow. I was pretty sure of that. You were both people he felt he could count on exploiting. After all, he had done it before."

"But he probably would have turned to Paul."

"That's why I arranged with Preston," he said, smiling, "to have Paul temporarily removed from the scene. Then I told Preston to get hold of Harrington and get over here."

"We were downstairs when Redwood arrived," Harrington added. "We followed him up here. Of course, we didn't know you were already here, Miss Foxworth. But you weren't really in danger, because Dobbs was already here too, in the bedroom."

"She certainly wasn't in any danger once Howard arrived," Dobbs said with satisfaction. "He was good, wasn't he, my dear?"

"He was wonderful," Dinah said, hugging me.

"I was just angry," I muttered.

"The lion protecting his cub," Dobbs stated with a flourish. "You see, dear boy, faced with real danger, instead of imagined ones, you acted with estimable courage. I'm very proud of you."

"And so am I . . . my lion," Dinah said, adding a kiss.

"I think we're in danger of becoming superfluous company, Myles. Besides, I can't stand mushy scenes," Dobbs said. "Let's you and me repair to a friendly neighborhood tavern. Then I'm off to New Jersey, where nothing exciting—or mushy—ever happens."

Harrington regarded us uncomfortably. "Um, yes," he said. "Just one thing, Miss Foxworth . . . I don't know how much of this affair Miller has told you. . . ." He looked at me crossly. "Or is going to tell you. But I would like to remind you that . . ."

"I understand, Mr. Harrington; my lips are sealed. I'm sure I wouldn't want to breach national security."

"I wasn't thinking of national security, Miss Foxworth. I was thinking of the publishing-luncheon circuit and the love of a good story. Most of this is going to come out sooner or later. It's just that . . . a good deal of the time . . . I sort of had egg on my face. I wouldn't want you to think that all our operations are quite so . . ." He struggled for the right word.

"Never fear," she said laughing.

We all laughed at Harrington's evident relief.

"And you, Mr. Dobbs," Dinah said, taking hold of his arm. "I'm sure we all have a great deal to thank you for. But I want to thank you . . . for myself . . . for being here when Redwood arrived."

She planted a big kiss on his cheek, and for the first time, I saw him completely nonplused.

"Come on, Harrington," he growled. "The mush is beginning."

We saw them to the door, locked it carefully behind them, and clutching each other, we stood in the entrance hall. After a few warm kisses, Dinah pulled herself away.

"And now," she said, "I'll get on with cooking dinner."

"Suits me," I said, following her into the living room. "And after dinner, we can just stay in and watch a good ballgame on television. Do you like to watch baseball?"

She stood in the kitchen doorway, adjusting an apron around her waist. "I didn't . . . until now," she said with a smile. "And if it turns out I still don't, we can always find something better to do."